An Introduction to Queueing Systems

An Introduction to Queueing Systems

Sanjay K. Bose
Indian Institute of Technology
Kanpur, India

Kluwer Academic / Plenum Publishers
New York, Boston, Dordrecht, London, Moscow

Library of Congress Cataloging-in-Publication Data

Bose, Sanjay K.
An introduction to queuing systems/Sanjay K. Bose.
p. cm.
Includes bibliographical references and index.
ISBN 0-306-46734-8
1. Queuing theory. I. Title.

QA274.8 .B67 2002
519.8′2—dc21

2001053983

ISBN 0-306-46734-8

233 Spring Street, New York, New York 10013

http://www.wkap.com

10 9 8 7 6 5 4 3 2 1

A C.I.P. record for this book is available from the Library of Congress

Printed in the United States of America

Preface

Queueing is an aspect of modern life that we encounter at every step in our daily activities. Whether it happens at the checkout counter in the supermarket or in accessing the Internet, the basic phenomenon of queueing arises whenever a shared facility needs to be accessed for service by a large number of jobs or customers. The study of queueing is important as it provides both a theoretical background to the kind of service that we may expect from such a facility and the way in which the facility itself may be designed to provide some specified grade of service to its customers.

Our study of queueing was basically motivated by its use in the study of communication systems and computer networks. The various computers, routers and switches in such a network may be modelled as individual queues. The whole system may itself be modelled as a queueing network providing the required service to the messages, packets or cells that need to be carried. Application of queueing theory provides the theoretical framework for the design and study of such networks. The purpose of this book is to support a course on queueing systems at the senior undergraduate or graduate levels. Such a course would then provide the theoretical background on which a subsequent course on the performance modelling and analysis of computer networks may be based. This is indeed the strategy adopted for teaching computer networks and their performance modelling and analysis in the Department of Electrical Engineering at I.I.T., Kanpur. The material of this book was originally provided as lecture notes to the students specialising in the area of telecommunications and networking. These students were required to go through a sequence of two courses, the first one on queueing and the second on computer networks. The first course on queueing and the associated lecture notes were expected to provide the theoretical background for the second course and its related topics.

The phenomenon of queueing also arises in operations research and industrial engineering as the facilities studied in these areas may also be modelled as either individual queues or queueing networks. This text will also be useful for senior students in this area. This book assumes that the student is familiar with the basics of probability theory and its applications. It also assumes that the students know and can apply results from the theory of transforms, especially Laplace Transforms and Z-Transforms. It is normally expected that students in electrical engineering or computer science would be familiar with this required theoretical background before they reach their senior undergraduate or graduate levels. Students in other areas may require additional mathematical training to acquire this knowledge. An appropriate mathematical course covering these topics may be a prerequisite for such students who may want to use this book for the study of queueing systems.

This book is basically concerned with the analysis of queueing systems. Some example scenarios are usually given for the systems being studied even though studies of potential queueing applications are not the primary objectives of this text. The analytical models have been carefully developed and presented. However, given the expected mathematical background of the students using this text, obvious steps in the analysis have usually been omitted and left as an exercise for the readers. If this book is used for students studying queueing at a level lower than what was suggested earlier, the analytical steps might be omitted to focus more on the results which are finally obtained. For such students, it may be useful to just cover the basic queueing models of Chapter 2 and the basics of queueing networks as given in Chapter 5. For students or practicing engineers interested only in applying exact or approximate queueing network models for solving their application problems, the algorithms provided in Chapters 5 and 6 are recommended.

Chapter 1 provides an introduction to queues and queueing systems. To illustrate the simplicity of basic queueing analysis, this chapter also presents the analysis of a simple single server queue. This has been done with the aid of some simplifying assumptions and is intended to show that a basic queueing analysis may indeed be very easily done.

Chapter 2 presents the equilibrium solutions of basic queues that may be analysed using birth-death models. These queues are mostly ones with Poisson arrival processes and exponential service time distributions. We also discuss Little's result, which is used extensively in this book and elsewhere for the study of queues and queueing networks. Queues with bulk arrivals and using the method of stages to approximately solve queues with non-exponential service time distributions have also been considered in this chapter.

Chapters 3 and 4 deal with the single server queue with Poisson arrivals and general service time distributions. While Chapter 3 focusses on the basic queue of this type and presents an extensive analysis of its important performance parameters, Chapter 4 discusses some important variations of this queue. These variations include queues with vacations of different types and various priority queueing models. It also presents the analysis of some basic discrete-time queues. (All the other queueing models considered in this book are continuous-time in nature.)

Chapters 5 and 6 of the book deal with queueing networks. Chapter 5 considers simple open and closed network models that have a product-form solution and can be analysed exactly, provided suitable simplifying assumptions are satisfied. Jackson's Theorem is introduced here for a variety of queueing networks. We also present the convolution algorithm and the mean value algorithm for the exact solution of simple closed queueing networks. Norton's theorem for reducing closed queueing networks has also been discussed in this chapter.

Chapter 6 focusses on various algorithms that have been proposed as approximate methods of analysis for more complex queueing networks. A variety of such algorithms have been presented for both open and closed networks. Fork/Join nodes and their analysis in open and closed queueing network have also been considered. For open networks, we present the GI/G/m approximation of QNA that gives exceptionally good results in most situations. Networks where some or all the queues are of finite capacity have also been considered. The different blocking models that may be used have been presented and discussed in detail in this chapter. Even though these networks with blocking cannot be solved in an exact fashion (except with very restrictive assumptions) the Maximum Entropy Method does provide a reasonably good approximate solution. This has been presented for both open and closed networks with repetitive service blocking of different types. We have also presented approximate algorithms to handle open and closed networks with transfer blocking and open networks with rejection blocking.

Exact analysis of queueing systems is often difficult. Even with simplifying assumptions and approximations, it may not always be possible to obtain an analytical solution. In such cases, one has to take recourse to simulations for the study of these systems. Chapter 7 discusses the simulation approach and provides some of the basic knowledge necessary to meaningfully study queueing systems using simulations.

This book may be used as a textbook for a course on basic queueing theory by limiting oneself to the study of single queues of various types using Chapters 2, 3 and 4. The material on queueing networks in Chapters 5 and 6 may then be omitted from such a course. Chapter 6 gives summary descriptions of various approximation techniques that may be used to solve complex queueing networks. These would also be useful for practicing engineers who need to solve more complex queueing based network models. We have implemented the algorithms described in Chapters 5 and 6 in a queueing analysis package *QNAT*. A beta version of this software is being publicly distributed and may be downloaded from the Internet. The download details for this package may be obtained from the author. It should be noted that *QNAT* allows both analysis and simulation of all the queueing network models described in Chapters 5 and 6. For simulations, *QNAT* uses the discrete event simulation approach described in Chapter 7.

Sanjay Kumar Bose

Acknowledgements

This book would not have been possible without the help and feedback of my students at the Indian Institute of Technology, Kanpur. Their inputs during my course on Queueing Systems are deeply appreciated. I would like to especially thank Animesh Kumar who painstakingly corrected the draft notes and made valuable suggestions on the contents of the text.

My faculty colleagues in I.I.T. Kanpur, N.T.U. Singapore and R.M.I.T. University Melbourne helped in deciding the content of the book and its presentation format. I would especially like to thank Prof. Les Berry and Prof. Richard Harris of RMIT for their suggestions and advice. Prof. Bill Henderson of the University of Adelaide had originally suggested that I offer a course on queueing as a prerequisite to my course on Computer Networks. Since this book evolved from the teaching of that course, I would like to thank Bill for motivating this book. I would also like to thank my graduate students M.N. Umesh, D.M. Bhaskar, and Hema Tahilramani for their valuable inputs to this text. The input received from Prof. D. Manjunath of I.I.T. Mumbai is also deeply appreciated.

The effort involved in writing the original set of lecture notes on which this book is based and the preparation of this text would not have been possible without the help and support of my wife Aditi and my daughter Atreyi. I hope they would forgive me now for the evenings I spent poring over the computer at home. This book is dedicated to them for their support and understanding. This book is also dedicated to Prof. Stephen S. Rappaport of S.U.N.Y., Stony Brook, as he is the one who, many years ago, taught me the basics of queueing that I would like to think I know.

Contents

List of Figures

An Introduction to Queueing Systems

Chapter 1

Introduction

Queueing systems are models of systems providing service. Such a model may represent any system where jobs or customers arrive looking for service of some kind and depart after such service has been provided. We can model systems of this type as either single queues or a system of interconnected queues forming a queueing network. These are the kinds of systems that are dealt with here and our objective is to describe the analytical techniques that may be applied to study their performance. An example of a simple queueing model has been shown in Figure 1.1. Such a model may be used to represent a typical queueing situation where jobs arrive, wait if all servers are busy, eventually get served by an available server and leave after the required service is obtained.

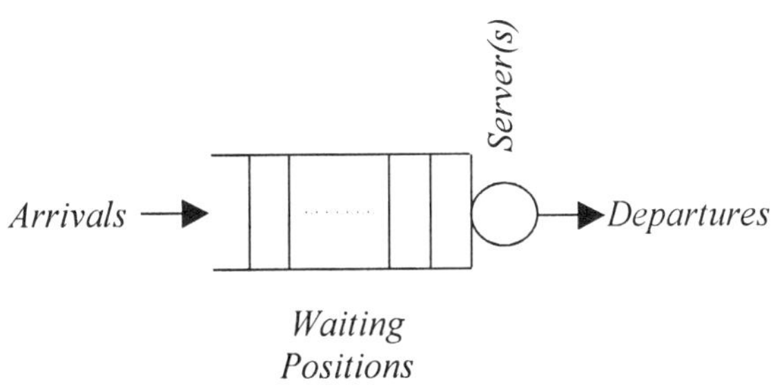

Figure 1.1. Model of a Simple Queue

1.1 Queueing Model Parameters

The analytical modelling of a queue would involve providing the input specifications describing the queue. These would include details on the arrival processes of jobs/customers to the queue, the service process at the servers of the queue, the number of servers and the number of buffers and waiting positions in the queue. For a system being modelled as a network of queues, details like the above will need to be provided for each of the individual queues. In addition, one would need to provide details on the way jobs are routed from one queue to another and the methods by which blocking, if any, is to be tackled.

The process of analysis of an individual queue aims at providing details on the kind of service a customer can expect as well as on how the queue itself is performing. The service parameters, which will be of interest to a customer looking for service at the queue, are the *queueing delay, total delay, number in the system* and the *number waiting in the queue*. Ideally, we would like to know the distribution of these quantities. However, if that is not possible (and it usually is not), then at least the mean values should be known. Note that for a new job entering the system, the mean values of all these service parameters should be low if satisfactory service is to be provided. These mean values are inter-related and may also be found if we can find the *equilibrium state distribution* of the number of jobs in the system under the given conditions. In systems with finite waiting capacities, an additional parameter of interest will be the mean and distribution of *blocked customers*, i.e. the number of customers leaving without service.

The server utilisation, mean and possibly its distribution, would be a quantity of interest not so much to the users of the queue as to the service provider. This is because a service provider would ideally like to maximise resource utilisation so as to maximise revenues, even though this may adversely affect user satisfaction by increasing delays and the number forced to wait in the queue. The nature of the departure process is also useful, if it can be found, as this would make it possible to analyse queueing networks where a part or whole of the output of one queue gets redirected to another.

If we make suitable simplifying assumptions, the analysis of queueing systems may be made quite simple, especially if we are only interested in the mean values of the performance measures. We have illustrated this in the next section. It may not always be possible to find the distributions of these parameters and even when these can be found the process may be a difficult or tedious one. Even when exact analysis is not possible or feasible, very effective approximation techniques may be used to provide good results. When even that cannot be done, or cannot be done easily enough, the only recourse left is to use simulations for studying the system.

1.2 A Simple Queueing Model

If we are not averse to making some crude simplifying assumptions, then analysis of simple queueing models may be easily done. (These assumptions will be justified later.) For example, consider the queue of Figure 1.1 with only one server and an infinite number of waiting positions. Let the arrival process of jobs be such that arrivals come at rate λ. Assume that when $\Delta t \to 0$, P{one arrival in time interval Δt}, the probability of one arrival in a time interval Δt, is $\lambda \Delta t$. Similarly, P{no arrivals in time interval Δt] is $1-\lambda \Delta t$ and P{more than one arrival in time interval Δt} will be negligibly small and may be considered to be zero. For the service process of the given server, we assume that the average service rate is μ (mean service time is $1/\mu$). Assume that when the server is working and when $\Delta t \to 0$, P{one departure from the system in time interval Δt] is $\mu \Delta t$, the P{no departures in time interval Δt] is $1-\mu \Delta t$ and P{more than one departure in time interval Δt] is zero.

For this queue, we need to describe the *system state* in some fashion. The system state at any time instant may be taken as the number in the system at that instant. Note that this will include both the number waiting in the queue and the customer currently in service, if any. Let $p_N(t)$ = P{system in state N at time t}. By ignoring terms with $(\Delta t)^2$ and higher order terms, the probability of the system state at time $t+\Delta t$ may then be found as

$$p_0(t+\Delta t) = p_0(t)[1-\lambda\Delta t] + p_1(t)\mu\Delta t \qquad N=0 \qquad (1.1)$$

$$p_N(t+\Delta t) = p_N(t)[1-\lambda\Delta t-\mu\Delta t] + p_{N-1}(t)\lambda\Delta t + p_{N+1}(t)\mu\Delta t \qquad N>0 \qquad (1.2)$$

subject to the normalisation condition that $\sum_{\forall i} p_i(t) = 1$ for all $t \geq 0$.

Taking the limits as $\Delta t \to 0$, and subject to the same normalisation, we get

$$\frac{dp_0(t)}{dt} = -\lambda p_0(t) + \mu p_1(t) \qquad N=0 \qquad (1.3)$$

$$\frac{dp_N(t)}{dt} = -(\lambda + \mu)p_N(t) + \lambda p_{N-1}(t) + \mu p_{N+1}(t) \qquad N>0 \tag{1.4}$$

These differential equations (along with the normalisation condition) may be used to get both the *transient* and the *equilibrium* solutions. For the transient solutions, the queue is assumed to start at time $t=0$ with some initial state K, i.e. $p_i(0)=\delta_{iK}$ and the differential equations are solved to obtain the state probabilities $p_i(t)$ $i=0, 1, 2.....\infty$ as a function of the time t. For the equilibrium solutions, the conditions invoked are

$$\frac{dp_i(t)}{dt} = 0 \text{ and } p_i(t) = p_i \text{ for } i=0, 1, 2.....\infty$$

For this, defining $\rho = \lambda/\mu$ *erlangs*, with $\rho < 1$ for stability, we get

$$\begin{aligned} p_1 &= \rho p_0 \\ p_{N+1} &= (1+\rho)p_N - \rho p_{N-1} = \rho p_N = \rho^{N+1} p_0 \qquad N \geq 1 \end{aligned} \tag{1.5}$$

Solving Eq. (1.5) with the normalisation condition $\Sigma\, p_i = 1$, we get the system state probabilities to be

$$p_i = \rho^i (1-\rho) \qquad i = 0, 1, \ldots\ldots, \infty \tag{1.6}$$

It should be noted that the summation in the normalisation condition would only have a finite value when $\rho<1$. This condition is therefore required for the queue to be stable. Note that once we know the equilibrium state probabilities, we can use these to compute various mean performance parameters of this *simple queue*, Some examples of these are given next.

(a) Mean Number in System, N

$$N = \sum_{i=0}^{\infty} i p_i = \sum_{i=0}^{\infty} i \rho^i (1-\rho) = \frac{\rho}{1-\rho} \tag{1.7}$$

(b) Mean Number Waiting in Queue (prior to service), N_q

$$N_q = \sum_{i=1}^{\infty} (i-1) p_i = \frac{\rho}{1-\rho} - (1-p_0) = \frac{\rho}{1-\rho} - \rho = \frac{\rho^2}{1-\rho} \tag{1.8}$$

(c) Mean Time Spent in System, W

To obtain this, we need to assume *first come first served* (FCFS) service even though the result obtained does actually hold for any other service discipline where "*the server is never idle while the system is non-empty*". Note that this equality only holds for the mean. The second moment and higher moments, as well as the actual distribution will be different for different service disciplines.

Assuming FCFS discipline, consider a job, which arrives to the system when it is in state k, i.e. there are k users already in the system of which one is currently being served. We need to make one more assumption before we can proceed further with this analysis. We will assume that the mean remaining service time needed to finish serving the customer currently being served, is still $1/\mu$. This is indeed justified if the service time distribution is *exponential* (with mean $1/\mu$), which is effectively what is being assumed here. (This is because of the memory-less property of the exponential distribution.) With these assumptions, we get

$$W = \sum_{k=0}^{\infty} \frac{(k+1)}{\mu} p_k = \frac{1}{\mu(1-\rho)} \tag{1.9}$$

(d) Mean Time Spent waiting in Queue, prior to service, W_q

It is easy to see that W_q will always be one mean service time less than the value of W obtained earlier from Eq. (1.9). Hence, we get

$$W_q = W - \frac{1}{\mu} = \frac{\rho}{\mu(1-\rho)} \tag{1.10}$$

Note that we can also obtain W_q directly if we know the probabilities of the system states. Using the same arguments and assumptions as in (c) above (for deriving W), we can see that a customer arrival which finds the system in state k will encounter a mean queueing delay of k/μ. This may also be used to obtain W_q directly as shown below.

$$W_q = \sum_{k=0}^{\infty} \frac{k}{\mu} p_k = \frac{\rho}{\mu(1-\rho)} \tag{1.11}$$

Note that this mean result will also be independent of the service discipline actually followed.

(e) Server Utilisation

This will be a parameter of interest to the service provider arranging for the server serving the jobs in the queue. Obviously, the service provider would like the server to be utilised as much as possible, even though that might lead to large queueing delays and a large number of jobs waiting for service in the queue.

The server is busy except when the system is empty, i.e. the system state is zero. Therefore, we can see that the utilisation of the server under equilibrium conditions will be $1-p_0 = \rho$. This quantity is indicative of how hard the server is actually working.

(f) Probability that an arriving customer has to wait for service

In a way, this parameter is indicative of the quality of service being provided by the queue. Obviously, customers will not be happy with the service provided if they have to wait for service every time they arrive at the queue. From the customers' point of view, a good system will be one where they immediately start getting served on arrival.

It is obvious that a customer will get served immediately on arrival, i.e. will not have to wait for service, if it sees the queue to be empty on arrival. The probability of this is simply p_0, the probability that the system is empty and will be given by $1-\rho$. Note that as the traffic ρ $(0<\rho<1)$ increases, the server utilisation improves at the cost of lower customer satisfaction with the quality of service being provided by the queue.

The *simple queue* described above is actually what is referred to as the $M/M/1/\infty$ queue, i.e. a queue with Poisson arrivals, exponentially distributed service times, single server and infinite waiting positions. (The notation used for representing the queue is called *Kendall's Notation* and will be described later.)

In general, we find that the analysis of a single server queue is attractive, because -

(a) It is generally more tractable than the analysis of a multi-server queue

(b) For a queue with C servers, bounding results may be obtained by considering a system with C parallel, single server queues where an arriving customer can join any of the queues randomly. The latter system will always be less efficient than the actual C server queue.

1.3 Some Basic Queueing Models

The approach given here is always applicable for simple queues where the arrival process may be modelled as a Poisson process (with exponentially distributed inter-arrival times) and the service times are exponentially distributed. This happens because of the memory-less property of the exponential distribution. This simple approach will not be applicable if these conditions are not met. We can easily extend this approach to analyse a variety of queueing situations. Some examples of these are given below.

Finite Buffer Queues

In this case, the system's state will be limited to a value K which is the maximum number of users who can be in the system, i.e. the waiting customers and the ones in service. Arrivals, which come when the system is full, are forced to leave without service as they are not allowed to enter the queue.

Multiple Servers

In this case, the service rate should be taken as $k\mu$ for the system state where k servers are serving

Variable Arrival Rate

This, for example, can be the case where the arrival rate is $(N-k)\lambda$ when the system is in state k. A model of this kind is frequently used to denote the arrival process from a system serving a population of finite number of users, i.e. N users. Here a user, who has generated a job that is currently in the system, does not generate a new one until the earlier job finishes service.

Queues with "Balking"

This, for example, could be a system where an arriving job checks the number, say k, presently in the system. It then decides to join the queue with a probability $e^{-\alpha k}$ or leaves without service with probability $(1-e^{-\alpha k})$, where α, $0<\alpha<1$, is an appropriately chosen constant. Several variations of this are also possible such as the case where the arriving user bases its decision for "balking" on either the number that it sees *waiting for service* in the queue or its estimated waiting time

1.4 A Summary of the Contents

In Chapter 2, we consider simple queueing models with Poisson arrival processes and exponential service times. Extension to a generalised service time distribution for single-server queues is considered in Chapter 3 and other issues like vacations, priorities, batch arrivals and simple discrete-time queues are considered in Chapter 4. Chapter 5 considers simple models of systems where queues are interconnected to form a queueing network. Approximate methods that can be used to handle more complicated queueing networks are discussed in Chapter 6. Chapter 7 ends this text with a description of simulation techniques and the issues that arise in simulation studies of individual queues and queueing networks.

Chapter 2

Basic Queueing Theory

Fundamentals of Analyzing Single Queues

As shown in the simple example of the previous chapter, the basic approach to the analysis of simple queueing models would begin by defining an appropriate system state for the queue. The analysis of the queue would then essentially be the study of the way this system state would evolve. The *transient solution* would be the solution obtained for this system state, given the various input parameters, and the initial conditions with which the queue starts operation. In this text, we are however interested in the performance analysis of the queue once equilibrium conditions have been reached. Analyses of some basic queues where the arrivals come from a Poisson process and the service times are exponentially distributed will be considered in this chapter. Before we consider such analysis, it would be useful to review some of the basics of the theory of *Markov Chains* and *Birth-Death Processes*. These are considered next. Further details on this may be found in [Fel65], [Kle75] or [Wol89].

2.1 Markov Processes and Markov Chains

Markov processes are memory-less in nature, which makes models using such processes easier to handle. For a stochastic process $X(t)$, this memory-less property, also sometimes called the *Markov Property*, states that for any choice of time instants t_i $i=1, \ldots\ldots, n$, we have

$$P\{X(t_{n+1})=x_{n+1} \mid X(t_n)=x_n \ \ldots\ldots\ X(t_1)=x_1\}=P\{X(t_{n+1})=x_{n+1} \mid X(t_n)=x_n\}$$

Note that when this property is satisfied, the state of the process/system at time instant t_{n+1} depends only on the state of the process/system at the previous instant t_n and not on any of the earlier time instants. Alternatively, one can say that a process is termed a *Markov Process* if, given the present state of the process, its future evolution is independent of the past of the process. This effectively implies a *one-step dependency feature* for the *Markov Process* where older values are forgotten. Restricted versions of this property leads to special cases, such as -

(a) Markov Chains over a Discrete State Space

(b) Discrete-Time and Continuous-Time Markov Processes and Markov Chains

If the random variables denoting the state of the process are discrete in nature, then these discrete random variables $\{X_n\}$ form a *Markov Chain* if the probability that the next state is x_{n+1} depends only on the current state x_n and not on any previous values. For the *Discrete Time* case, state changes are pre-ordained to occur only at the integer points $0, 1, 2, \ldots\ldots, n$ (that is at the time points $t_0, t_1, t_2, \ldots\ldots, t_n$). However, for the *Continuous Time* case, state changes may occur anywhere in time.

In the analysis of simple queues, the state of the queue may be represented by a single random variable $X(t)$ which takes on integer values $\{i, i=0, 1\ldots\ldots\}$ and the corresponding process may be treated as a *Continuous Time Markov Chain*. Such a chain is referred to as a *Homogenous Markov Chain* if the probability $P\{X_{n+1}=j \mid X_n=i\}$ is the same regardless of n. In that case, one can write the transition probability p_{ij} of going from state i to state j as $p_{ij} = P\{X_{n+1}=j \mid X_n=i\}\ \forall n$. It should be noted that for a homogenous Markov chain, the transition probabilities depend only on the terminal states (i.e. the initial state i and the final state j) but do not depend on when the transition $(i \rightarrow j)$ actually occurs.

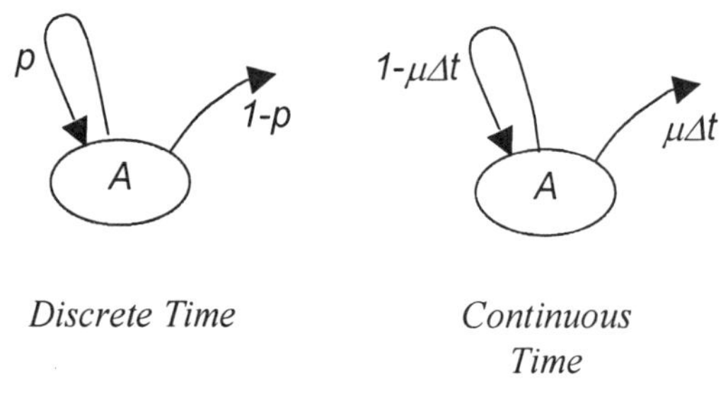

Figure 2.1. State Transition from State A

Consider the state transition for a *homogenous, discrete-time, Markov Chain* as shown in Figure 2.1 and let $P\{X_{n+1}=A|X_n=A\}=p$ and $P\{X_{n+1}\neq A|X_n=A\}=(1-p)$. Then for this Markov Chain, we can show that

P{system stays in state A for N time units | given that the system is currently in state A} $= p^N$

and

P{system stays in state A for N time units before exiting from state A} $= p^N(1-p)$

Note that the above result gives a *Geometric Distribution* that would be memory less in nature.

Similar results may also be obtained for the *Continuous Time* case. In this case, assume that the system is in state A at time t and that the departure rate from state A is μ. We consider time $t+T$ to find the probability that the system stays in state A for an additional time T, given that the current time is t. We can then show that as $\Delta t \rightarrow 0$

P{system in state A for time T | system currently in state A}

$$= (1-\mu\Delta t)^{T/\Delta t} \rightarrow e^{-\mu T}$$

Note that this is the *cumulative distribution function* (cdf) of the exponential distribution. Based on this and the earlier result, we can make the important observation that "*In a homogenous Markov Chain, the distribution of time spent in a state is (a) Geometric for discrete time or (b) Exponential for continuous time*". The implication of this statement is that in a *homogenous Markov Chain*, either discrete or continuous, the form of the distribution of time spent in a particular state of the system is fixed - i.e. it cannot be arbitrarily chosen. The distribution is also memory-less in nature. This implies that if we examine the system at a particular instant of time and find the system in a particular state (say state A), then the additional amount of time the system will spend in this state (before exiting to another state) will not depend on how much time has already been spent in this state. This condition gets relaxed if one considers *Semi-Markov Processes*. In these processes, the distribution of time spent in a state can have an arbitrary distribution but the one-step memory feature of the Markovian property is retained. We will find processes of this type useful in some of our analyses.

A sequence of random variables $\{S_n\}$ is a *Random Walk* if -

$S_0 = 0$ (assume process starts at origin, without loss of generality)
and $S_n = S_{n-1} + X_n$ for $n=1, 2, \ldots\ldots\ldots, \infty$
where $X_1, X_2, \ldots$.etc. are independent, identically distributed (i.i.d.) random variables. A random process related to this is the *Renewal Process*. This is related to a random walk except that our interest here lies in counting the number of transitions that take place as a function of time. Let the state at time t be given by the number of transitions in $(0, t)$ where t_i corresponds to the time of the ith transition and $(t_i - t_{i-1})$ $\forall i$ are i.i.d. random variables. Let $X_i = (t_i - t_{i-1})$ $\forall i$ denote this set of i.i.d. random variables. Subject only to the conditions that they are independent and have identical distributions, the random variables $\{X_i\}$ can have any distribution. Note that this would actually correspond to a *Semi-Markov Process*.

In particular, consider the special case where $X_i = \tau$ with an exponential distribution given by $\lambda e^{-\lambda\tau}$. It can then be shown that the renewal (counting) process for this is the *Poisson Process* such that

$$P\{N \text{ transitions in an interval } T\} = (\lambda T)^N \frac{e^{-\lambda T}}{n!} \tag{2.1}$$

corresponding to the *Poisson Distribution*. To show this, let $p_k(t)$ be the number of transitions up to time t. (In the queueing scenario, this corresponds to a queue where arrivals occur but no service is provided - the arrivals are merely stored. In this case, $p_k(t)$ will be the probability that the queue has k arrivals up to time t, given that queue was empty at time $t=0$). Therefore

$$p_k(t+\Delta t) = p_k(t)[1-\lambda\Delta t] + p_{k-1}(t)\lambda\Delta t \qquad \text{for } k=1, 2, \ldots\ldots.$$

and with $\Delta t \rightarrow 0$, we get

$$\frac{dp_0(t)}{dt} = -\lambda p_0(t)$$

$$\frac{dp_k(t)}{dt} = -\lambda[p_k(t) - p_{k-1}(t)] \qquad k = 1, 2, \ldots\ldots..$$

Solving these for $p_0(0)=1$ and $p_k(0)=0$ for $k=1, 2, \ldots\ldots$, we get Eq. (2.1).

2.1.1 Discrete Time Markov Chains

We review this in more detail, as most of our analysis would be based on this kind of a Markov chain. As defined earlier, the sequence of random variables $X_1, X_2, \ldots\ldots$ forms a Markov Chain if for all n $(n=1, 2, \ldots\ldots)$ and all possible values of the random variables, we have that

$$P\{X_n=j \mid X_1=i_1 \ldots\ldots X_{n-1}=i_{n-1}\}=P\{X_n=j \mid X_{n-1}=i_{n-1}\}$$

Note once again that this property really illustrates a *one-step memory* where the next state depends only on the current state and $P\{X_n=j \mid X_{n-1}=i_{n-1}\}$ gives the *one step transition probability* of going from state i_{n-1} at the $(n-1)^{th}$ step to state j at the nth step.

If the discrete-time Markov Chain is *homogenous* in nature, then the state transition probability $p_{ij}= P\{X_n=j \mid X_{n-1}=i\}$ will also be independent of n, i.e. the instant when the transition actually occurs. For the homogenous, discrete-time Markov Chain, one can also define the *m-step state transition probability* as

$$p_{ij}^{(m)} = P\{X_{n+m} = j \mid X_n = i\} = \sum_{\forall k} p_{ik}^{(m-1)} p_{kj} \quad m=2, 3, \ldots. \tag{2.2}$$

A Markov Chain is said to be *Irreducible* if every state can be reached from every other state in a *finite* number of steps. This implies that k exists such that $p_{ij}^{(k)} \neq 0$ for $\forall i, j$. Note that if a Markov Chain is *not irreducible*, then -

(a) it may have one or more *absorbing states* - i.e. states from which the process cannot move to any of the other states, or,

(b) it may have a subset of states A from where one cannot move to states in A^C, i.e. states outside A

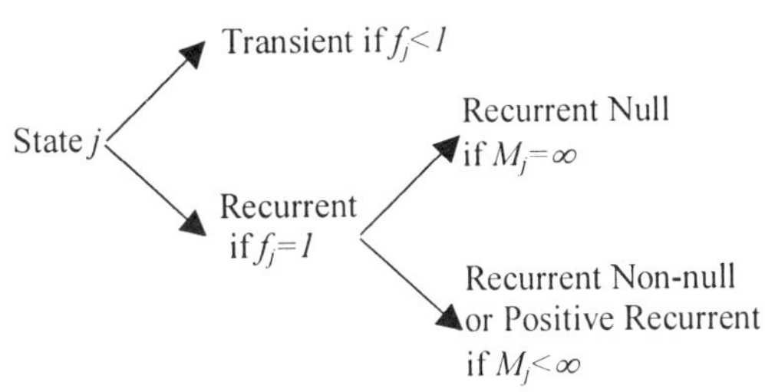

Figure 2.2. Classification of States for a Markov Chain

The states of a Markov chain may be classified further as being either *Recurrent* or *Transient*. Recurrent states may be further divided into

Recurrent Null or *Recurrent Non-null*. This classification has been illustrated in Figure 2.2.

A state is considered to be *recurrent*, if it will occur some time whereas a state would be *transient* if it may not occur at all. Consider the probability $f_j^{(n)}$ of the system returning to state j in exactly n steps after leaving state j. Note that trivially, $f_j^{(1)}=p_{jj}$ and that we can derive the probability f_j of the system returning to state j some time after leaving state j to be

$$f_j = \sum_{n=1}^{\infty} f_j^{(n)} \tag{2.3}$$

From the definition of f_j, we can say that that the state j will be transient if $f_j<1$, whereas it will be recurrent if $f_j=1$. For a state j, which is recurrent, i.e. can occur again, its *mean recurrence time* M_j may be defined as

$$\textit{Mean Recurrence Time for state } j = M_j = \sum_{n=1}^{\infty} n f_j^{(n)}$$

A state j is *periodic* with respect to α $(\alpha > 1)$, if the only possible steps in which state j may occur are $\alpha, 2\alpha, 3\alpha$................ In that case, the recurrence time for state j has *period* α. Note that state j is said to be aperiodic if $\alpha=1$.

A *Recurrent State* is said to be *Recurrent Null* if the mean recurrence time for that state tends to infinity, i.e. $M_j=\infty$. A *Recurrent State* is said to be *Recurrent Non-null* (also called *Positive Recurrent*) if the mean recurrence time for that state is finite, i.e. $M_j<\infty$. A *Recurrent State* is said to be *Ergodic* if it is both positive-recurrent and aperiodic. This implies that it is possible to come back to that state in any given number of steps and that such a return will always occur with finite mean recurrence times. An *Ergodic Markov Chain* will have all its states as ergodic. Note that an *Aperiodic, Irreducible, Markov Chain with a finite number of states* will always be ergodic. The property of ergodicity is important as it implies that regardless of the initial state $p_j^{(0)}$, the equilibrium state probabilities obtained for the system, i.e. $\lim_{n\to\infty}\{p_j^{(n)}\}\to p_j$, would still be the same. Note that the probability of the system being in state j in the nth step will be given as

$$P\{\textit{system in state } j \textit{ in the } n^{th} \textit{ step}\} = p_j^{(n)} = P\{X_n=j\}$$

The following results then hold [Kle75], [Wol89] for an *Irreducible, Markov Chain* with $p_j^{(n)}$ as defined above -

[A] The states of an Irreducible Markov Chain are either all transient, or all recurrent null or all recurrent positive. If the chain is periodic, then all states have the same period α.

[B] In an irreducible, aperiodic, homogenous Markov Chain, the limiting probabilities $p_j=\lim_{n\to\infty}\{p_j^{(n)}\}$ always exist and these are independent of the initial state probability distribution and either -

(a) All states are transient, or all states are recurrent null - in this case, the state probabilities p_j's are zero for all states and no stationary state distribution will exist.

or (b) All states are recurrent positive - in this case a *stationary distribution* giving the equilibrium state probabilities exists and is given by $p_j=1/M_j \ \forall j$. These state probabilities may also be calculated by solving the following simultaneous equations

$$\sum_{\forall i} p_i = 1 \quad \text{or} \quad \boldsymbol{pe=1} \qquad \text{(Normalisation Condition)} \qquad (2.4)$$

$$p_j = \sum_{\forall i} p_i p_{ij} \quad \forall j \quad \text{or} \quad \boldsymbol{p=pP} \qquad \text{(Balance Condition)} \qquad (2.5)$$

where $\boldsymbol{p}=(p_0, p_1, \ldots\ldots\ldots\ldots p_n)$ is the state probability vector and $\boldsymbol{P}=\{p_{ij}\}$ is the state transition probability matrix. Note that $\boldsymbol{e}^T=(1, \ldots\ldots, 1)$ is the unit vector.

2.2 Birth-Death Processes

A *Birth-Death Process* is a special type of discrete-time or continuous time Markov Chain with the restriction that at each step, the state transitions, if any, can occur only between neighbouring states. Without loss of generality, we can assume that the State Space of the process is the set of integers. This may be done as this is basically just a way of labelling the states. If the process is a Birth-Death process and if the current state X_n is i, then the above condition implies that the next state X_{n+1} can only be $i-1$, i or $i+1$. This states that in successive time instants, the state either stays unchanged or has a unit increment or decrement. A variation of this will be a *Pure Birth Process* where decrements are not allowed. One may also have a *Pure Death Process* where increments are not allowed, i.e. the system starts from an initial state and decrements to zero as time goes on.

We consider here the Birth-Death Process as a *Continuous-Time Markov Chain*; the discrete-time case may be similarly handled. Note that since a continuous-time process is being considered, we need to focus on changes in the process over time interval Δt as $\Delta t\to 0$. Let λ_k be the birth rate in state k.

Note that when the system is in state k this birth rate is the rate at which the system state can increase by one. Similarly, let μ_k be the death rate in state k, i.e. the rate at which the system state can decrease by one in state k.

In the queueing theory context, the birth rate would typically correspond to the rate of arrivals of new jobs or customers to the queue whereas the death rate would be the rate at which customers/jobs leave the queue after completion of service. Note that since we have made the implied assumption that the corresponding Markov Chain is homogenous, the birth and death rates *do not depend on time*; they may however be state dependent and will then depend on the current state of the system, i.e. state k.

Considering transitions over a time interval Δt as $\Delta t \to 0$, we get that

$P\{state\ k\ to\ state\ k+1\ in\ time\ \Delta t\} = \lambda_k(\Delta t)$

$P\{state\ k\ to\ state\ k\text{-}1\ in\ time\ \Delta t\} = \mu_k(\Delta t)$

$P\{state\ k\ to\ state\ k\ in\ time\ \Delta t\} = 1 - (\lambda_k + \mu_k)(\Delta t)$

$P\{other\ transitions\ in\ \Delta t\} = 0$

Let $X(t)$ be the number in the system at time t, i.e. (total births - total deaths) up to time t. Let $p_k(t)$ be the probability of finding the system in state k at time t, i.e. $p_k(t) = P\{X(t) = k\}$. For $X(t)=0, 1, \ldots\ldots\ldots\ \infty$, we can then write the following equations for the state transitions between the time instant t and the time instant $t+\Delta t$.

$$p_0(t+\Delta t) = p_0(t)[1-\lambda_0 \Delta t] + p_1(t)\mu_1 \Delta t$$

$$p_k(t+\Delta t) = p_k(t)[1-(\lambda_k+\mu_k)\Delta t] + p_{k-1}(t)\lambda_{k-1}\Delta t + p_{k+1}(t)\mu_{k+1}\Delta t$$

$$with\ \sum_{k=0}^{\infty} p_k(t) = 1 \qquad (Normalisation\ Condition)$$

These correspond to the state transition diagram shown in Figure 2.3.

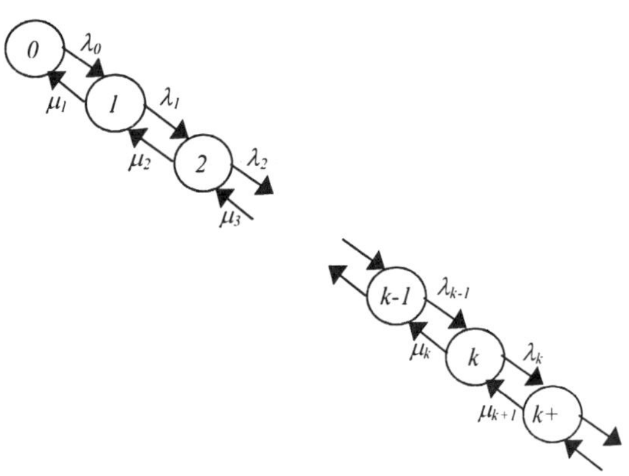

Figure 2.3. State Transition Diagram for a Birth-Death Process

With $\Delta t \to 0$, this yields the following

$$\frac{dp_0(t)}{dt} = -\lambda_0 p_0(t) + \mu_1 p_1(t)$$

$$\frac{dp_k(t)}{dt} = -(\lambda_k + \mu_k) p_k(t) + \lambda_{k-1} p_{k-1}(t) + \mu_{k+1} p_{k+1}(t) \tag{2.6}$$

$$\sum_{k=0}^{\infty} p_k(t) = 1$$

which may be solved with the proper initial conditions to get the required solutions $\{p_k(t)\}$ $k=0, 1, \ldots\ldots \infty$ for the state distribution of the system as a function of the time t. Note that the last equation in the above comes from the normalising condition which states that the sum of the probability of all the states will be unity at any instant of time t. Other than this equation, the other equations in Eq. (2.6) may alternatively be derived using a *flow-based approach*. This is illustrated in the following for state k, $k \geq 1$ - the considerations for the case $k=0$ are similar and may be easily applied.

(a) Flow into state k = $\lambda_{k-1}p_{k-1}(t) + \mu_{k+1}p_{k+1}(t)$
(b) Flow out of state k = $(\lambda_k+\mu_k)p_k(t)$

(c) $\dfrac{dp_k(t)}{dt}$ = Flow into state k - Flow out of state k

$$= \lambda_{k-1}p_{k-1}(t) + \mu_{k+1}p_{k+1}(t) - (\lambda_k+\mu_k)p_k(t)$$

A *Flow-Based Approach* will typically also be the easiest way of solving some of the simpler queueing problems. This approach would typically be usable if the arrival process (generating *births*) is a Poisson process and the service process (generating *deaths*) has exponentially distributed service times. In that case the typical approach would be to -

(a) Draw the state transition diagram
(b) Draw closed boundaries and equate flows across this boundary. If the closed boundary encloses state k then we get

$$\lambda_{k-1} p_{k-1} + \mu_{k+1} p_{k+1} = (\lambda_k + \mu_k) p_k$$

as the desired equation for state k. Note that the dependence on t has been dropped because equilibrium conditions are being examined.

Closed boundaries may also be drawn and used. For example, we can draw the closed boundary as a straight line in-between the states (*k -1*) and *k*, which is actually *closed at infinity*. In this case, the flow balance equation that we will get will be $\lambda_{k-1}p_{k--1}=\mu_k p_k$

It can be shown that the system of equations would be equivalent regardless of the type of closed boundary chosen to write the equilibrium flow conditions. However, it is possible that one set of equations will be easier to manipulate and use than another set - something that can be seen in the above.

(c) Solve the equations in (b) to obtain the equilibrium state probability distribution

Flow Balance Equations may also be generalised as follows. For any irreducible, aperiodic, homogenous, Markov Chain at equilibrium, we can write the following as *flow balance across the boundary enclosing state j*

$$\sum_{i \neq j} p_i p_{ij} = p_j \sum_{i \neq j} p_{ji} \qquad \textit{Global Balance Equation}$$

Note that the LHS of the above corresponds to the total flow from states, other than state *j*, to state *j* and the RHS corresponds to the flow from state *j* to other states. This is referred to as the *Global Balance Equation* for the *Birth-Death Process* and would lead to the equations $\lambda_{k-1}p_{k-1}+\mu_{k+1}p_{k+1} = (\lambda_k+\mu_k)p_k \ \forall k$ given earlier.

In addition to the Global Balance Equation (which always hold for a system in equilibrium), *Detailed Balance Equations* may also hold if the flow from state *i* to state *j* equals the flow from state *j* to state *i*. For the Birth-Death Process considered here, these will indeed hold and we can write.

Flow from state *i* to state *j* = $p_i p_{ij}$
Flow from state *j* to state *i* =$p_j p_{ji}$

$$p_i p_{ij} = p_j p_{ji} \qquad \textit{Detailed Balance Equation}$$

Note that for the system considered earlier, the *Detailed Balance Equations* would lead to the equations $\lambda_k p_k = \mu_{k+1}p_{k+1} \ \forall k$ for the process. These could also been derived earlier with a suitable choice of the closed boundary.

Since we have

$$\lambda_k p_k = \mu_{k+1} p_{k+1} \qquad k=0, 1, 2, \ ...... \ \infty$$

we get that

$$p_{k+1} = \frac{\lambda_k}{\mu_{k+1}} p_k \qquad k=0, 1, 2, \ ...... \ \infty$$

This may be simplified to yield the following probabilities for the system states.

$$p_k = p_0 \left[\prod_{i=0}^{k-1} \frac{\lambda_i}{\mu_{i+1}} \right] \qquad k=1, 2, \ ...... \ \infty \tag{2.7}$$

The form of the expression for p_k in Eq. (2.7) is important. This is an example of a *Product Form Solution* as it consists of a continued-product of terms multiplied by a normalising constant. This form of expressions for the state probabilities at equilibrium would be commonly encountered in many queueing situations. Applying the normalising condition

$$\sum_{i=0}^{\infty} p_i = 1$$

to the expression for p_k, we will obtain p_0, the probability of the system being empty as

$$p_0 = \frac{1}{1 + \sum_{k=1}^{\infty} \prod_{i=0}^{k-1} \frac{\lambda_i}{\mu_{i+1}}} \tag{2.8}$$

It should be noted that in our derivation above, we have not mentioned the conditions under which such steady state (i.e. equilibrium) solutions may be obtained. We have actually tacitly assumed that the conditions are such that equilibrium exists and hence the steady state solutions given in Eq. (2.7) and Eq. (2.8) will hold. This needs to be more formally stated, as this requires the Markov Chain to be *Ergodic* in nature. To define the general condition for the equilibrium solution to exist, we need to define two variables α and β as given below.

$$\alpha = \sum_{k=0}^{\infty}\left[\prod_{i=0}^{k-1}\frac{\lambda_i}{\mu_{i+1}}\right] \quad \text{and} \quad \beta = \sum_{k=0}^{\infty}\frac{1}{\lambda_k \prod_{i=0}^{k-1}\frac{\lambda_i}{\mu_{i+1}}}$$

Having obtained α and β, we can then observe the following for the states of the Markov Chain -

(a) All states are *transient*, if and only if $\alpha = \infty, \beta < \infty$
(b) All states *recurrent null,* if and only if $\alpha = \infty, \beta = \infty$
(c) All states *ergodic*, if and only if $\alpha < \infty, \beta = \infty$

We actually need the condition (c) in the above, for our solution, as equilibrium (steady state) exists only when all states of the Markov Chain are ergodic. Ideally, the condition $\alpha < \infty$, $\beta = \infty$ should be verified to hold before we can attempt to obtain the steady state solutions for the process. In practice, some equivalent but simpler statements may be made regarding when the system will have steady state solutions. One simple test is to ensure that the system is such that $p_0 > 0$ does indeed exist. This has an interesting implication as it essentially states that *the system will have an equilibrium solution only if it is such that there is a finite probability that it will sometimes be empty,* i.e. will be in state *0*. Note that by finding p_0 in our solution approach, we essentially did apply this condition implicitly. It can also be shown that the *ergodic condition* is met *if k_0 exists such that $(\lambda_k/\mu_k) < 1$ for all $k \geq k_0$*. This condition for ergodicity basically affirms that beyond a certain point in the state space of the process, the rate of new arrivals for the states must fall to values that are always lower than the corresponding service rate for these states. It is intuitively obvious that if this condition is fulfilled then the system state cannot keep growing and will have to come down sometime. This would therefore lead to stable and equilibrium conditions.

2.3 Kendall's Notation for Queues

This is a useful way to represent different types of queues in a compact and easily understood fashion. Kendall's Notation describes the nature of the arrival process to the queue, the nature of the service process (in terms of the service time), the number of servers, the maximum number of jobs that may be in the system and some basic queueing disciplines. The notation has been considerably extended to allow it to represent a wide variety of queues. We give here a very basic description of this notation.

Following this representation, a queue is represented by a sequence *A/B/C/D/E* with the following meaning attached to the letters *A* to *E*.

A This symbolically represents the nature of the arrival process to the queue. Special letters are used to symbolise the nature of the *inter-arrival time distribution*. Some of the important ones are

- M Exponentially distributed inter-arrival times (Poisson Process)
- D Deterministic (fixed) inter-arrival times
- E_k Erlang distribution of order k for inter-arrival times
- H_k Hyper-exponential distribution of order k for inter-arrival times
- G General (any!) distribution for the inter-arrival times

B This symbolically represents the nature of the service time distribution for the customers getting served in the queue. The same letters as the ones above are used to describe the nature of the *service time distribution*

C Number of servers in the queue

D Maximum Number of jobs/customers that can be there in the system. This includes both the ones currently being served and the ones waiting for service. Note that the default is *infinity* (∞) which is assumed when this is omitted.

E Queueing Discipline such as -

- FCFS First Come First Served
- LCFS Last Come First Served
- SIRO Service In Random Order

This may also be omitted if the queue is FCFS in nature (default).

Examples:

M/M/1 or M/M/1/∞	Poisson Arrivals, Exponential Service Time Distribution, Single Server, Infinite Number of Waiting Positions
M/E_2/2/K	Poisson Arrivals, Erlangian of order-2 Service Time Distribution, Two Servers, Maximum Number K in system (waiting and in service)

G/M/2	Generalised Arrivals, Exponential Service Time Distribution, 2 Servers, Infinite Number of Waiting Positions

2.4 Little's Result

Consider a queue that has a mean arrival rate of λ (actually entering the queue). If N is the mean number in system and W is the mean delay (i.e. the mean time spent in the queueing system by a customer including both its waiting time and service time), then Little's Result states that

$$N=\lambda W \qquad \textit{(Little's Result)} \qquad (2.9)$$

will hold under equilibrium conditions. Note that Little's Result will hold for virtually all kinds of queueing systems under very general conditions. It will certainly hold for all the queues that we will consider here. Note that if we define N_q as the mean number waiting in queue (prior to service) and W_q as the mean waiting time in queue (prior to service), then these are also similarly related.

$$N_q=\lambda W_q \qquad (2.10)$$

A graphical approach demonstrating Liitle's Result is given based on Figure 2.4 assuming FCFS service and a queue, which is initially empty.

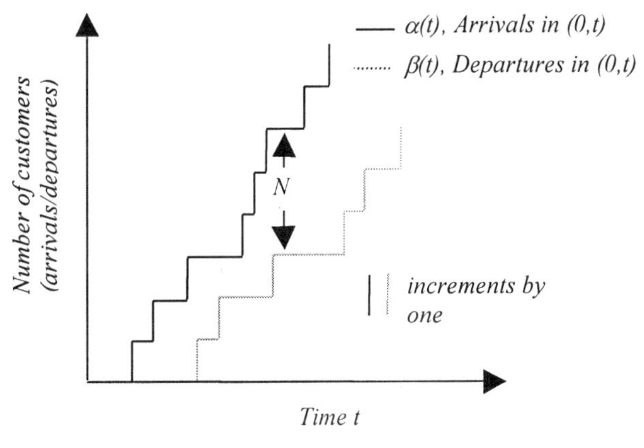

Figure 2.4. Graphical Verification of Little's Result

Let $\alpha(t)$ be the number of arrivals in $(0, t)$ and let $\beta(t)$ be the number of departures in $(0, t)$. Then $\alpha(t) - \beta(t)$ will be the number in the system at time t.

$$Area(t) = \text{area between } \alpha(t) \text{ and } \beta(t) \text{ upto time } t = \int_0^t [\alpha(t) - \beta(t)]dt$$

Note that $Area(t)$ will also be the *total time spent in system by all arrivals which enter the system by time t.* We can also see that the following will also hold.

$$\text{Average arrival rate in the interval } (0, t) = \lambda_t = \frac{\alpha(t)}{t}$$

$$\text{Average arrival rate (actual)} = \lambda = \lim_{t\to\infty} \lambda_t = \lim_{t\to\infty} \frac{\alpha(t)}{t}$$

The average time W_t spent in the system by all arrivals who have come in $(0,t)$ will be $Area(t)/\alpha(t)$. The average number of jobs N_t in the system observed over the interval $(0,t)$ is $Area(t)/t$. We can then write

$$N_t = \frac{Area(t)}{t} = \frac{\alpha(t)}{t}\frac{Area(t)}{\alpha(t)} = \lambda_t W_t \tag{2.11}$$

The average time W spent in system by an arrival will be approximately given by the limit $\lim_{t\to\infty} W_t$. This neglects the minor errors caused by the fact that the jobs in the system at time t would not have finished service at time t. As t becomes large, the contribution of this error will decrease to zero. Taking the limits of Eq. (2.11) as $t\to\infty$, gives the result of Eq. (2.9)

Note that this graphical illustration of Little's Result makes the simplifying assumptions that (a) the service is FCFS and (b) the system is originally empty. These assumptions are actually not needed for Little's Result to hold. In general, Little's Result will hold for virtually all queueing system in equilibrium except under some very unusual conditions [BeG92]. It will certainly hold for all the systems considered in this text.

2.5 Equilibrium Solutions for M/M/-/- Queues

In this section, we study the equilibrium solutions for queues where the arrivals come from a Poisson Process (i.e. exponentially distributed inter-arrival times) and the service required by the jobs/customers in the queue have an exponentially distributed service time distribution. The Markov Chain describing these queues are Birth-Death Processes and the equilibrium state distributions for these queues may be directly obtained by appropriately applying the *product form solution* given by Eq. (2.7) and Eq. (2.8). We will subsequently use this to obtain the equilibrium state distributions for a wide variety of M/M/-/- queues. Similar examples may also be seen in [BeG92], [Kle75] and [Wol89].

However, before we proceed to do this, we need to look a little more closely at the way we have obtained the expression for p_k given by the product form expression. Note that this is obtained by defining the queue's state to be the total number in the system and using this to draw its state transition diagram. The equilibrium solution is then obtained by appropriately applying flow balance to this. This in effect gives us the *time-averaged* (and hence ergodic, steady state) distribution of the state probabilities. If an outside observer were to observe the state of the queue, the state probability p_k essentially gives the fraction of time he/she would find the system to be in state k. It should be noted that these results on the state probabilities do not necessarily hold for other types of observations that one might make on this system. Examples of these would be -

(a) Observe the system states only at the time instants when new arrivals occur
(b) Observe the system states only at the time instants when a job/customer departs from the system
(c) Observe the system at a fixed time of day or with a fixed periodicity

Consider the situation in (a) when we observe the system states only at the instants when a new arrival occurs and want to find the probability distribution of the states observed at those instants. Note that this would correspond to the state probability distribution as observed by the arrivals coming to the queue. For this, the following important result is useful.

2.5.1 PASTA - Poisson Arrivals See Time Averages

The PASTA property claims that for queues where the arrival process is Poisson (i.e. M/-/-/- type queues), the state probability distribution as seen by a new arrival (coming from the arrival process) to the queue would be the same as the time-averaged (i.e. ergodic, steady state) state probability distribution.

To prove the PASTA property for Poisson arrivals, let $p_k(t)$ be the probability that the system is in state k at time t and let $q_k(t)$ be the probability that an arrival at time t finds the system in state k. Let $A(t, t+\Delta t)$ be the event of an arrival in the time interval $(t, t+\Delta t)$ and let $N(t)$ be the actual number in the system at time t. We then have

$$
\begin{aligned}
q_k(t) &= \lim_{\Delta t \to 0} P\{N(t) = k \mid A(t, t+\Delta t)\} \\
&= \lim_{\Delta t \to 0} \frac{P\{A(t, t+\Delta t) \mid N(t) = k\} P\{N(t) = k\}}{P\{A(t, t+\Delta t)\}}
\end{aligned} \tag{2.12}
$$

However, since the arrivals come from a Poisson Process with exponentially distributed service times, we can use the memory-less property of the process to claim that the number of arrivals $A(t, t+\Delta t)$ in the time interval $(t, t+\Delta t)$ cannot depend on the state of the system $N(t)$ at time t. Therefore, $P\{A(t, t+\Delta t) \mid N(t) = k\}$ would be equal to $P\{A(t, t+\Delta t)\}$. Hence

$$
q_k(t) = \lim_{\Delta t \to 0} P\{N(t) = k\} = p_k(t)
$$

as required by the PASTA property. Note that PASTA will not hold if the arrival rate depends on the state of the system. Obviously, it will also not hold for non-Poisson type of arrival processes. The usefulness of PASTA comes from the fact that it is sometimes relatively easier to obtain averages of performance parameters (like N, N_q, W, W_q) from the point of view of a newly arriving customer. If PASTA holds for the system then the results obtained from this may also be interpreted to hold for ergodic time averages.

2.5.2 M/M/1 (or M/M/1/∞) Queue

This is a single server queue with infinite number of buffers where the arrival process is Poisson (i.e. exponentially distributed inter-arrival times) and service times are exponentially distributed. Note that this is really the queue considered in a simplified fashion in Section 1.2. For this case, we have

$$
\lambda_k = \lambda \qquad \forall k
$$

$$
\begin{aligned}
\mu_k &= 0 \qquad && k = 0 \\
&= \mu \qquad && k = 1, 2, 3, \ldots\ldots
\end{aligned}
$$

We define $\rho=\lambda/\mu$ as the traffic offered to the queue in units of *erlangs*. Note that in this case, if $\rho<1$, then the parameters α and β of Section 2.2 are such that $\alpha<\infty$ and $\beta=\infty$. This implies that an equilibrium state probability distribution for the queue will exist if $\rho<1$, i.e. $\lambda<\mu$. This solution may then be found using Eq. (2.7) and Eq. (2.8) to be the same as given in Eq. (1.6). We find that in this case $p_0 = (1-\rho) > 0$ for $\lambda<\mu$. This then also satisfies the other way in which we had stated the condition for the existence of equilibrium state probability distributions for a Birth-Death Process (i.e. that the system must be such that its probability of being empty at equilibrium must be non-zero). Actually, we need to find p_0 by applying the normalising condition. The reader can verify that finding p_0 in this way is only possible if $\lambda<\mu$ or $\rho<1$ as otherwise the infinite sum cannot be evaluated.

The mean number N in the system and the mean number N_q waiting in queue may be found directly using the state probabilities as shown in Section 1.2. We have illustrated there how the mean time W spent in the system by an arriving customer may be found. Note that this derivation actually implies use of the PASTA principle as it uses the ergodic state distribution to be also the distribution as seen by an arriving customer. In addition, as mentioned in Section 1.2, the memory-less property of the exponential distribution is also used. The derivation used for this assumes an FCFS queue. However, the same mean result will apply for any queueing discipline where the server is not allowed to idle as long as there are jobs waiting for service in the queue. In particular, the mean time W spent in the system will be the same even if the service discipline is LCFS or SIRO. The same comments hold for the derivation of the mean time W_q spent waiting in queue (prior to service) by an arriving customer.

The distributions of the delays may also be found and are considered later in Section 2.6 for the FCFS discipline. In Chapter 4, we obtain the delay distributions for a general M/G/1 queue for both the FCFS and LCFS disciplines. These may be used to obtain the corresponding results for the simpler M/M/1 case. In general, it should be noted that although the means are the same, different disciplines would have different distributions, and different second and higher moments. Specifically, the FCFS discipline has the smallest variance and LCFS the highest, i.e. the spread of the delay distribution is least for FCFS and highest for LCFS, even though both have the same mean values.

Actually, of the four parameters calculated above, one really needs to calculate only one, as the other three may be found from that. For this, one can use the following relations -

1. $W = W_q + \mu^{-1}$ from mean service time considerations
2. $N = \lambda W$ from Little's result
3. $N_q = \lambda W_q$ from Little's result

2.5.3 M/M/1/∞ Queue with Discouraged Arrivals

For this case, we have

$$\lambda_k = \frac{\lambda}{k+1} \qquad \forall k$$

$$\begin{aligned} \mu_k &= 0 \qquad && k = 0 \\ &= \mu && k = 1, 2, 3, \ldots\ldots \end{aligned}$$

We refer to this type of arrival process as *discouraged arrivals* as in this case, the mean arrival rate progressively decreases as the state of the system goes up. This is as if we are discouraging new arrivals to this queue, when there are more people waiting in the queue for service. We will derive the usual performance parameters for this queue but the following two points should be kept in mind.

(a) PASTA will not be applicable here as the arrival process is *not Poisson* in nature

and (b) For applying Little's result, we need to use λ_{eff} which would have to be calculated as

$$\lambda_{eff} = \sum_k \lambda_k p_k = \sum_k \frac{\lambda}{k+1} p_k \tag{2.13}$$

after we find the equilibrium state probabilities $\{p_k\}$.

Since this is a still a Birth-Death Process with (state dependent) Poisson arrivals and exponentially distributed service times, the product-form solution of Eq. (2.7) and Eq. (2.8) may be applied to find the equilibrium state probability distribution. This will be given by

$$p_k = p_0 \prod_{i=0}^{k-1} \frac{\lambda}{\mu(i+1)} = p_0 \left(\frac{\lambda}{\mu}\right)^k \frac{1}{k!} \qquad \text{for } k = 1, 2, 3, \ldots\ldots \tag{2.14}$$

$$p_0 = \exp(-\frac{\lambda}{\mu}) \tag{2.15}$$

Note that we can find that the condition for ergodicity for this case to be $\lambda/\mu < \infty$. This will be the condition that needs to be satisfied for equilibrium conditions to exist. For a queue in equilibrium, the system performance parameters for this case may be found as follows.

Mean number in system $$N = \sum_{k=0}^{\infty} kp_k = \frac{\lambda}{\mu}$$

Average arrival rate to system $$\lambda_{eff} = \sum_{k=0}^{\infty} \lambda_k p_k = \mu\left[1 - \exp(-\frac{\lambda}{\mu})\right]$$

Using these and Little's result, we get the mean time W spent in system by a customer (including both the waiting time and the time to get service) as

$$W = \frac{N}{\lambda_{eff}} = \frac{\lambda}{\mu^2\left[1 - \exp(-\frac{\lambda}{\mu})\right]}$$

In this case, we can compute W directly as well. However, for that, we need to find the probability distribution of the queue as seen by an arriving customer. Note that this will be different from the equilibrium state probability distribution obtained above because PASTA will not hold for this queue. Let π_r = P{arriving customer sees r in system (before joining the system)} and let ΔE be the event of an arrival in $(t, t+\Delta t)$, where E_i is the event of the system being in state i

Then $$\pi_r = P\{E_r \mid \Delta E\} = \frac{P\{E_r\}P\{\Delta E \mid E_r\}}{P\{\Delta E\}} = \frac{P\{E_r\}P\{\Delta E \mid E_r\}}{\sum_{i=0}^{\infty} P\{E_i\}P\{\Delta E \mid E_i\}}, \quad \Delta t \to 0$$

Using $P\{E_i\} = p_i = e^{-\lambda/\mu}\left(\frac{\lambda}{\mu}\right)^i \frac{1}{i!}$ and $P\{\Delta E \mid E_i\} = \frac{\lambda \Delta t}{i+1}$ and simplifying,

we get $\pi_r = \left(\frac{\lambda}{\mu}\right)^{r+1} \frac{1}{(r+1)!}\left(\frac{e^{-\lambda/\mu}}{1 - e^{-\lambda/\mu}}\right)$ as the required probability.

This may now be used to directly compute W as follows

$$W = \sum_{k=0}^{\infty} \frac{k+1}{\mu} \pi_k = \frac{\lambda}{\mu^2 (1 - e^{-\lambda/\mu})}$$

giving the same expression as that obtained earlier using Little's Result

2.5.4 M/M/m/∞ Queue (*m* servers, infinite number of waiting positions)

For this case, we have the average arrival rates given by

$$\lambda_k = \lambda \qquad \forall k$$

and the average service rates as

$$\begin{aligned} \mu_k &= k\mu & 0 \le k \le (m-1) \\ &= m\mu & k \ge m \end{aligned} \qquad \text{or} \qquad \mu_k = \min(k\mu, m\mu)$$

Defining $\rho=\lambda/\mu$, we get that equilibrium conditions exist if $\rho<m$ and that the equilibrium state probability distribution $\{p_k\}$ may be found using the product form solution of Eq. (2.7) and the normalising condition of Eq. (2.8) to be

$$\begin{aligned} p_k &= p_0 \frac{\rho^k}{k!} & \textit{for} \quad k \le m \\ &= p_0 \frac{\rho^k}{m!\,m^{k-m}} & \textit{for} \quad k > m \end{aligned} \tag{2.16}$$

$$p_0 = \left(\sum_{k=0}^{m-1} \frac{\rho^k}{k!} + \frac{m\rho^m}{m!(m-\rho)} \right)^{-1} \tag{2.17}$$

For this system, an important performance measure that is usually mentioned is the probability of queueing. This may be found as

P{queueing} = P{arriving customer has to wait for service}

$$\text{or } P\{\text{queueing}\} = \sum_{k=m}^{\infty} p_k = C(m,\rho) = p_0 \frac{m\rho^m}{m!(m-\rho)} \tag{2.18}$$

Note that $C(m,\rho)$ defined by Eq. (2.18) is a function that may usually be found in tables and is referred to as the *Erlang-C Formula*. This is because this queue was first analysed by Erlang as a model of a telephone exchange with m outgoing lines (i.e. m servers) where callers who arrive to find all lines busy, wait until a line becomes free. This corresponds to a queue with an infinite number of waiting positions, since there is no limit on the number of people who may wait for a line. Note that this model of a telephone exchange could be one where there is an operator who connects callers to outgoing lines and asks the callers to hold if all lines are busy.

2.5.5 M/M/m/m Queue (m server loss system, no waiting)

For this case, we have the average arrival rates given by

$$\begin{aligned} \lambda_k &= \lambda && k < m \\ &= 0 && \textit{otherwise} \quad (\textit{Blocking or Loss Condition}) \end{aligned}$$

and the average service rates as

$$\begin{aligned} \mu_k &= k\mu && 0 \le k \le m \\ &= 0 && \textit{otherwise} \end{aligned}$$

Unlike the previous case, jobs enter this system only when a free server is available. If all servers are busy, the arriving customer does not enter the queue and leaves without service. This is a more realistic model for a telephone exchange where a caller gets one of the m outgoing lines only if one such line is free. If all the lines are busy, the caller is forced to leave without service.

Note that this is a loss system. If the probability of blocking is given as P_B, then the mean arrival rate actually entering the queue will only be $\lambda(1-P_B)$. This should be the λ_{eff} to be used as the λ in Little's result, if that is applied to calculate performance parameters. This system will always be ergodic as the system state is limited to $\{0, 1, ...m\}$. Hence, steady state will always exist and equilibrium state probability distributions may be found for all values of m, λ and μ. Using the product form expressions of Eqs. (2.7) and (2.8), and defining $\rho=\lambda/\mu$, we can find the state probabilities.

$$p_k = p_0 \frac{\rho^k}{k!} \qquad for \quad k \le m$$
$$= 0 \qquad otherwise \tag{2.19}$$

$$p_0 = \frac{1}{\sum_{k=0}^{m} \frac{\rho^k}{k!}} \tag{2.20}$$

The *Blocking Probability B(m,ρ)* for this queue is defined as the probability that an arrival finds all servers busy and leaves without service. This is given by

$$B(m,\rho) = p_0 \frac{\rho^m}{m!} \qquad \text{(with } \rho = \frac{\lambda}{\mu}\text{)} \tag{2.21}$$

The blocking probability *B(m, ρ)* is also a tabulated function that can be found and is commonly referred to as the *Erlang-B Formula* or *Erlang Blocking Probability*. It can also be easily computed based on the following recursion which may be derived from Eq. (2.21)

$$B(0,\rho) = 1 \qquad B(m,\rho) = \frac{\frac{\rho B(m-1,\rho)}{m}}{1 + \frac{\rho B(m-1,\rho)}{m}} \tag{2.22}$$

2.5.6 M/M/1/K Queue (single server queue with K-1 waiting positions)

In this queue, the amount of waiting space available is finite. Jobs/Customers who come when the system is full are not allowed to enter the system and have to leave without service. For this case, we have the average arrival rates given by

$$\lambda_k = \lambda \qquad k < K$$
$$= 0 \qquad otherwise \quad (Blocking \;\; or \; Loss \; Condition)$$

and the average service rates as

$$\mu_k = \mu \qquad k \le K$$
$$= 0 \qquad otherwise$$

We define $\rho=\lambda/\mu$, and note that in this case the states are restricted *(i.e. $k=0,1,2,.....,K$)* and cannot grow to infinity. This system will always be ergodic and steady state will always exist. The equilibrium state probability distribution may be found using Eqs. (2.7) and (2.8) and will be given by

$$p_k = p_0 \rho^k \qquad for \quad k \le K \tag{2.23}$$
$$= 0 \qquad otherwise$$

$$p_0 = \frac{(1-\rho)}{(1-\rho^{K+1})} \tag{2.24}$$

Note that once the equilibrium state probability distribution is known, the typical mean performance parameters N, N_q, W and W_q may be found - other higher moments, like variances etc., may also be calculated.

2.5.7 M/M/1/-/K Queue *(*single server, infinite number of waiting positions, finite customer population *K*)

Actually, since the size of the customer population is limited to K, the number of waiting positions in the queue need not be more than K-1. This is because customers arrive from the (finite) customer population to the queue and rejoin the customer population only after service is completed. For this case, we have the average arrival rates given by

$$\lambda_k = \lambda(K-k) \qquad k < K$$
$$= 0 \qquad otherwise \quad (Blocking\ \ or\ Loss\ Condition)$$

and the average service rates as

$$\mu_k = \mu \qquad k \le K$$
$$= 0 \qquad otherwise$$

Note that in this case also the states are restricted (i.e. $k=0,1,2,.....,K$) and cannot grow to infinity. This system will always be ergodic and steady state will always exist. Defining $\rho=\lambda/\mu$, the equilibrium state probability distribution may be found using Eqs. (2.7) and (2.8) and will be given by

$$p_k = p_0 \rho^k \frac{K!}{(K-k)!} \qquad k=1,\ 2,\ \ldots.,\ K \tag{2.25}$$

$$p_0 = \frac{1}{\sum_{k=0}^{K} \rho^k \frac{K!}{(K-k)!}} \tag{2.26}$$

2.5.8 M/M/∞ /-/K Queue (infinite servers, finite customer population)

This is really the same system as the one considered earlier in Section 2.5.7, except that we now assume that there are an infinite number of servers available in the queue. Therefore, an arriving job will always find a server available. For this case also, we have the average arrival rates given by

$$\begin{aligned} \lambda_k &= \lambda(K-k) && k<K \\ &= 0 && \textit{otherwise} \quad (\textit{Blocking or Loss Condition}) \end{aligned}$$

The average service rates will now be

$$\begin{aligned} \mu_k &= k\mu && k \le K \\ &= 0 && \textit{otherwise} \end{aligned}$$

Note that in this case also the states are restricted ($k=0,1,2,\ldots..,K$) and cannot grow to infinity. This system will always be ergodic and steady state will always exist. Defining $\rho=\lambda/\mu$, the equilibrium state probability distribution may be found using Eqs. (2.7) and (2.8) and will be given by

$$p_k = p_0 \rho^k \frac{K!}{(K-k)!k!} = p_0 \rho^k \binom{K}{k} \qquad k=1,\ 2,\ \ldots\ldots,\ K \tag{2.27}$$

$$p_0 = \frac{1}{\sum_{k=0}^{K} \rho^k \dfrac{K!}{k!(K-k)!}} = \frac{1}{(1+\rho)^K} \tag{2.28}$$

In this case, the mean number in the system N may also be easily computed and may be shown to be $K\rho\ /(1+\rho)$. We can also find λ_{eff}, the *effective arrival rate* to the queue as

$$\lambda_{eff} = \sum_{k=0}^{K} \lambda_k p_k = K\lambda \frac{1}{1+\rho} \tag{2.29}$$

Using Eq. (2.29) and the value of N, we can get the other mean parameters as $W=\mu^{-1}$, $W_q=0$ *and* $N_q=0$

2.6 Delay Analysis for FCFS M/M/1/∞ and M/M/m/∞ Queues

The analysis for obtaining the delay distributions of the M/M/1/∞ and the M/M/m/∞ queues for the FCFS service disciplines are given next. It should be noted that the delay distribution will depend on the nature of the service discipline, even though the mean delays will be the same regardless of the service discipline.

2.6.1 Delay Analysis for a FCFS M/M/1/∞ Queue

Let Q be the (random) queueing delay (not counting the service time) encountered by an arrival to this queue and let $f_Q(t)$ and $F_Q(t)$ be its probability density function (pdf) and cumulative distribution function (cdf), respectively. Similarly, let W be the random variable corresponding to the total delay (waiting and service time) and let $f_W(t)$ and $F_W(t)$ be its probability density function and cumulative distribution function respectively. Let $L_Q(s)$ and $L_W(s)$ be the respective Laplace Transforms (L.T.) of the pdf's of the queueing delay and total delay. Assume that the mean arrival rate to the queue is λ and the mean service time is $1/\mu$. The service time (r.v. T) has an exponential distribution with probability density function $\mu e^{-\mu t}$ $t \geq 0$ with its L.T. given by $\mu/(s+\mu)$. Note that $W=Q+T$ and that Q will be independent of T (i.e. $Q \perp T$). Therefore, we have

$$L_W(s) = L_Q(s)\frac{\mu}{s+\mu} \qquad (\text{or} \quad f_W(t) = f_Q(t) * [\mu e^{-\mu t}]) \tag{2.30}$$

This implies that knowing the distribution (pdf, cdf or L.T.) of either of the two random variables W and Q will allow the distribution of the other to be computed. In the following, we will first derive the distribution of Q and then use that to derive the distribution of W.

Consider a particular arrival of interest and the queueing delay Q that it will encounter. For this,

$F_Q(t)$ = P{queueing delay $\leq t$}
= P{queueing time=0} + [$\Sigma_{n\geq 1}$ P{queueing time$\leq t$ | arrival found n jobs in system}]p_n

The probability of n service completions in a time interval $(x, x+dx)$ is given by an *Erlang-n Distribution* given by

$$\text{P}\{n \text{ service completions in } (x, x+dx)\} = \left(\frac{\mu(\mu x)^{n-1}}{(n-1)!}e^{-\mu x}\right)dx \quad \text{for } x\geq 0$$

Note that these follow from the fact that the sum of n independent, identically distributed (iid) random variables with an exponential distribution will have an Erlang-n type distribution. Moreover, since PASTA is applicable, the probability that an arriving customer will find n customers in the system will be $(1-\rho)\rho^n$ from Eqs. (2.14) and (2.15). Using these, we get

$$\begin{aligned} F_Q(t) &= (1-\rho) + (1-\rho)\sum_{n=1}^{\infty}\rho^n \int_{x=0}^{t}\frac{\mu(\mu x)^{n-1}}{(n-1)!}e^{-\mu x}dx \\ &= (1-\rho) + (1-\rho)\rho\int_0^t \mu e^{-\mu x}\sum_{n=1}^{\infty}\frac{(\mu x\rho)^{n-1}}{(n-1)!}dx \qquad t\geq 0 \qquad (2.31) \\ &= (1-\rho) + (1-\rho)\rho\int_0^t \mu e^{-\mu x(1-\rho)}dx \\ &= (1-\rho) + \rho(1-e^{-\mu t(1-\rho)}) \end{aligned}$$

and $$f_Q(t) = \frac{dF_Q(t)}{dt} = \delta(t)(1-\rho) + \lambda(1-\rho)e^{-\mu t(1-\rho)} \qquad t\geq 0 \qquad (2.32)$$

Using (2.30), we can then get

$$f_W(t) = (1-\rho)\mu e^{-\mu t} + \lambda(1-\rho)\mu \int_0^t e^{-\mu(1-\rho)(t-x)} e^{-\mu x} dx \quad \text{for } t \geq 0$$

which may be simplified to

$$f_W(t) = (\mu - \lambda)e^{-(\mu-\lambda)t} \quad \text{for } t \geq 0$$

to get the desired probability density function of the total time W spent in the system by an arriving job.

2.6.2 Delay Analysis for the FCFS M/M/m/∞ Queue

Using the same notation as for the previous M/M/1/∞ case, consider the queueing delay Q and its cumulative distribution function $F_Q(t)$ as defined earlier. In the expression for $F_Q(t)$, using PASTA, the individual terms may be expressed as

$$\text{P\{queueing time=0\}} = \sum_{n=0}^{m-1} p_n = p_0 \sum_{n=0}^{m-1} \frac{\rho^n}{n!}$$

P{queueing time $\leq t$ | arrival found n in system}=P{$(n\text{-}m+1)$ service completions in $(0,t)$} $= \displaystyle\int_0^t \frac{m\mu(m\mu x)^{n-m}}{(n-m)!} e^{-m\mu x} dx$

Therefore, we can obtain $F_Q(t)$ as

$$F_Q(t) = p_0 \sum_{n=0}^{m-1} \frac{\rho^n}{n!} + p_0 \sum_{n=m}^{\infty} \frac{\rho^n}{m!\, m^{n-m}} \int_0^t \frac{m\mu(m\mu x)^{n-m}}{(n-m)!} e^{-m\mu x} dx \qquad (2.33)$$

Simplifying this, by summing the first term and interchanging the order of the summation and integration in the second, we get

$$f_Q(t) = \left\{1 - p_0\left[\frac{m\rho^m}{m!(m-\rho)}\right]\right\}\delta(t) + \left[\frac{\mu p_0 \rho^m e^{-\mu(m-\rho)t}}{(m-1)!}\right]u(t) \qquad (2.34)$$

where $u(t)$ is the unit step function and $\delta(t)$ is the delta function. Using Eq. (2.30), we can also obtain the probability density function $f_W(t)$ as

$$f_W(t) = \left\{1 - p_0\left[\frac{m\rho^m}{m!(m-\rho)}\right]\right\}\mu e^{-\mu t} - \left[\frac{\mu p_0 \rho^m [e^{-\mu(m-\rho)t} - e^{-\mu t}]}{(m-1)!(1-m-\rho)}\right] \quad (2.35)$$

Once the delay distributions $f_W(t)$ and/or $f_Q(t)$ (or the corresponding cdf's or Laplace Transforms) are known, we can calculate the other moments that we might want for these random variables. Specifically, the mean system delay W and queueing delay W_Q may be found from this. Using Little's Result, the mean number in the system N and the mean number waiting in queue N_Q may also be found for the queue in equilibrium.

Note that since PASTA is applicable to this queue (the arrival process is Poisson), N and N_Q will also be the average numbers (in system and waiting in queue, respectively) that an arriving customer will see. It would be interesting to find the distribution of the number in the system that will be seen by a departing customer - this would correspond to the number in the system that a customer who has just left the queue will see, looking back into the system. For this, consider a user of interest, who spends time t in the system before departure. Let N^* be the number in the system that it will see left behind when it departs with probability $p_n{}^*=P\{N^*=n\}$ for $N^*=0, 1,\ldots, \infty$. This situation is illustrated in Figure 2.5.

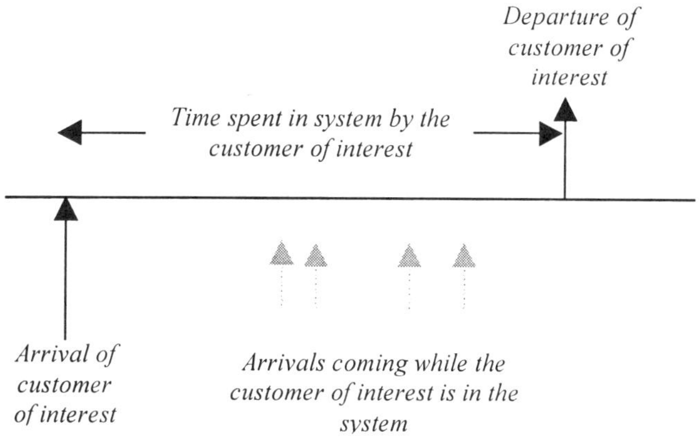

Figure 2.5. Arrival/Departure of Customer of Interest from a FCFS M/M/1Queue

From the above figure, it becomes evident that N^* will be equal to the number of arrivals (from a Poisson process with average rate λ) that enter the system in the time interval t, i.e. while the customer of interest is in the system. The *Generating Function* $G^*(z)$ for N^* will then be given by

$$\begin{aligned} G^*(z) &= \sum_{n=0}^{\infty} z^n p_n^* = \sum_{n=0}^{\infty} z^n \int_{t=0}^{\infty} \frac{(\lambda t)^n}{n!} e^{-\lambda t} f_W(t)dt \\ &= \int_0^{\infty} e^{-\lambda t(1-z)} f_W(t)dt = L_W(\lambda - \lambda z) \end{aligned} \tag{2.36}$$

Since $L_W(s)$ is known (i.e. it is the Laplace Transform of $f_W(t)$, the Generating Function $G^*(z)$ for N^* may be found. This can then be inverted to find the probability distribution of the number in the system as seen by a departing customer. The mean number $E\{N^*\}$ in the system as seen by a departing customer may be directly found using

$$E\{N^*\} = \left.\frac{dG^*(z)}{dz}\right|_{z=1} = -\lambda \left.\frac{dL_W(s)}{ds}\right|_{s=0} = \lambda W \tag{2.37}$$

which turns out to be the same as the value N obtained earlier in Section 2.5.2.

It is also important to note that the result $G(z)=L_W(\lambda-\lambda z)$ really says something which would be found generally useful. One way of viewing this result is to say *"if we observe Poisson arrivals at average rate λ over a random time interval T with* probability density function $f_T(t)$ *and L.T.* $L_T(s)$, *then the generating function of the number arriving in T will be given by* $L_T(\lambda-\lambda z)$*"*. This result is one we will use often.

2.7 Departure Process from a M/M/m/∞ Queue

The departure process from a queue would also be useful to characterise. Note that if a network of queues is being considered then the departure process of a queue would be the arrival process of one or more downstream queues. In such a system, we would certainly need to know the nature of the departure process from the first queue, in order to be able to analyse the behaviour of the downstream queue(s). *Burke's Theorem* provides the results necessary to look at the departure process of a M/M/m/∞ (which will include M/M/1 as well) queue. We present (without proof) the three statements associated with Burke's Theorem.

***Burke's Theorem*:** The following three statements hold for the departure process from a M/M/m/∞ queue -

[A] The departure process from a M/M/m/∞ queue is Poisson in nature.
[B] For a M/M/m/∞ queue, at each time t, the number of customers in the system is independent of the sequence of departure times prior to t.
[C] For a M/M/m/∞ FCFS queue, given a customer departure at time t, the arrival time of this customer is independent of the departure process prior to t.

Statement [A] from Burke's Theorem will actually be very useful to analyse networks of queues where the output of one queue feeds one or more queues. Consider the example shown in Figure 2.6.

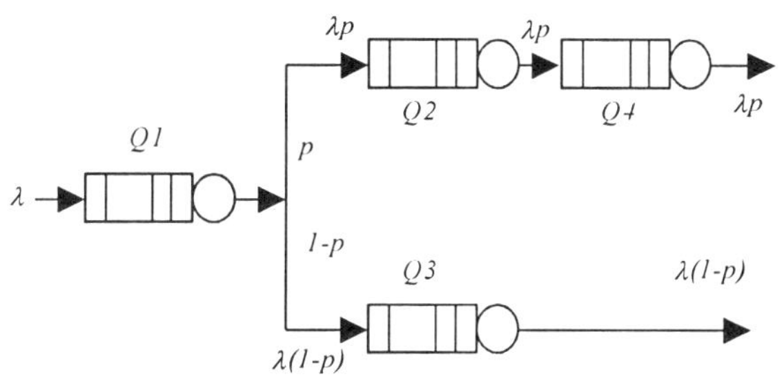

Figure 2.6. A Network of M/M/m Queues with Probabilistic Routing

In Figure 2.6 the arrival to *Q1* is from a Poisson Process with rate λ. The output process of *Q1* (by Burke's Theorem) is then also Poisson with rate λ. Note that flow equilibrium will ensure that the flow rate entering a queue at equilibrium would also be equal to the flow rate leaving the queue. The jobs leaving *Q1*, randomly decide with probability p to go to *Q2* and with probability $(1-p)$ to go to *Q3*. We use the result that splitting a Poisson flow randomly gives rise to flows that are also Poisson. Using this, the flows entering *Q2* and *Q3* will also be Poisson with average rates λp and $\lambda(1-p)$, respectively. The flow leaving *Q2* and entering/leaving *Q4* is also Poisson with rate λp and the flow entering and leaving *Q3* is Poisson with rate $\lambda(1-p)$. Note that the equilibrium state probabilities of each of these queues may be found as each of their respective arrival processes are also Poisson in nature with known rates. Note that we could not have done this without the application of Burke's Theorem as we would then be unable to claim that the arrival processes of *Q2*, *Q3* and *Q4* are also Poisson in nature. Knowing these, we can apply the results of Eqs. (2.16)-(2.18) to obtain the performance of each of the queues in the network.

Statement [B] of Burke's Theorem is somewhat counter-intuitive in nature. If one observes departures from a queue and finds a lot of jobs leaving it, then "lay" philosophy might say that the queue would be a busy one and we should find the system state to be high. However, Burke's Theorem shows that such a conclusion would be wrong as one cannot really say anything about the state of the queue by merely observing the sequence of departures from it. Statement [C] takes this further by stating that if we observe a customer departing at time t, then even if we are observing the departure process until time t, we cannot use this information to say anything about the time when that customer would have entered the system.

2.8 Time Reversibility Property of Irreducible, Aperiodic Markov Chains

Consider a discrete time, irreducible, aperiodic Markov Chain X_1, X_2, ..., X_{n-1}, X_n, X_{n+1}, ... for which the transition probabilities are given to be $\{p_{ij}\}$. These, for example, may arise in the study of M/M/m type queues. The transition probabilities may be used to obtain the equilibrium state probability distribution $\{p_i\}$ for this chain.

We can also consider the same chain backwards in time, i.e. the chain ......X_{n+1}, X_n, ..., X_3, X_2, X_1. This would also be a Markov Chain since we can write

$$
\begin{aligned}
&P\{X_m = j \mid X_{m+1} = i, X_{m+2} = i_2, \ldots\ldots, X_{m+k} = i_k\} \\
&= \frac{P\{X_m = j, X_{m+1} = i, X_{m+2} = i_2, \ldots\ldots, X_{m+k} = i_k\}}{P\{X_{m+1} = i, X_{m+2} = i_2, \ldots\ldots, X_{m+k} = i_k\}} \\
&= \frac{P\{X_m = j, X_{m+1} = i\} P\{X_{m+2} = i_2, \ldots\ldots, X_{m+k} = i_k \mid X_m = j, X_{m+1} = i\}}{P\{X_{m+1} = i\} P\{X_{m+2} = i_2, \ldots\ldots, X_{m+k} = i_k \mid X_{m+1} = i\}}
\end{aligned}
$$

Since the original (forward) chain is Markovian, the conditioning on both $\{X_m=j\}$ and $\{X_{m+1}=i\}$ in the numerator is equivalent to just conditioning by $\{X_{m+1}=i\}$ in the denominator. Therefore, we get

$$P\{X_m = j \mid X_{m+1} = i, X_{m+2} = i_2, \ldots\ldots, X_{m+k} = i_k\}$$
$$= \frac{P\{X_m = j, X_{m+1} = i\}}{P\{X_{m+1} = i\}} = \frac{P\{X_m = j\}P\{X_{m+1} = i \mid X_m = j\}}{P\{X_{m+1} = i\}}$$
$$= \frac{p_j p_{ji}}{p_i} = p_{ij}^*$$

From the above, we can state that for the backward chain, conditioned on the state at *(m+1)*, the state at *m* is independent of the states at *(m+2), (m+3),* ....etc. and that the backward chain has the transition probability $p_{ij}^* = p_j p_{ji}/p_i$. The Markov Chain is considered to be *time reversible* for the special case where $p_{ij}^* = p_{ij}$ $\forall\, i, j$. The following properties of the *reverse chain* may be demonstrated -

1. The reversed chain is also irreducible and aperiodic like the forward chain
2. The reversed chain has the same stationary state distribution as the forward chain
3. The chain is *time reversible* only if the detailed balance equation $p_i p_{ij} = p_j p_{ji}$ holds for $\forall\, i, j \geq 0$

Similar ideas of reversibility may also be applied to *Irreducible Continuous Time Markov Chains*.

2.9 The Method of Stages for Solving a M/-/1/∞ FCFS Queue

This method is useful to analyse queues where the service time is not exponentially distributed but may be represented (at least approximately) as a sum of independent, exponentially distributed random variables. These variables may not be identical but should be independent. (As we will show later, other more complex cases may also be handled by this method.

To illustrate this method, consider a single server queue where the service time is a random variable consisting of the sum of two, independent, exponentially distributed random variables with means $1/\mu_1$ and $1/\mu_2$. The arrival process is Poisson with rate λ and we assume that the queue has an infinite buffer. The operation of this queue may be illustrated as in Figure 2.7. In this queue, a customer starting service, first enters *Stage 1*, where it gets served for an exponentially distributed random time with mean $1/\mu_1$. On completion of this, it enters *Stage 2* for an exponentially distributed random time with mean $1/\mu_2$ and then departs from the system. New customers enter *Stage 1* only after the previous customer has left the system, i.e. left *Stage 2*.

The overall service time would thus be the sum of the random service times at the two stages.

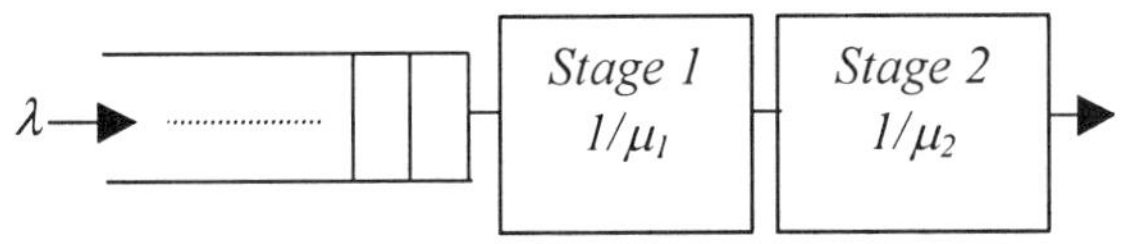

Figure 2.7. Single Sever Queue with Two Stages of Service

We can denote the state of the system in Figure 2.7 as *(n, j)* where *n* is the total number of customers in the system where the customer currently being served is at Stage *j*, $n=0,1,\ldots\ldots,\infty$, $j=1, 2$. We use *(0,0)* to denote the special case when the system is empty. The State Transition Diagram of this system will be as shown in Figure 2.8.

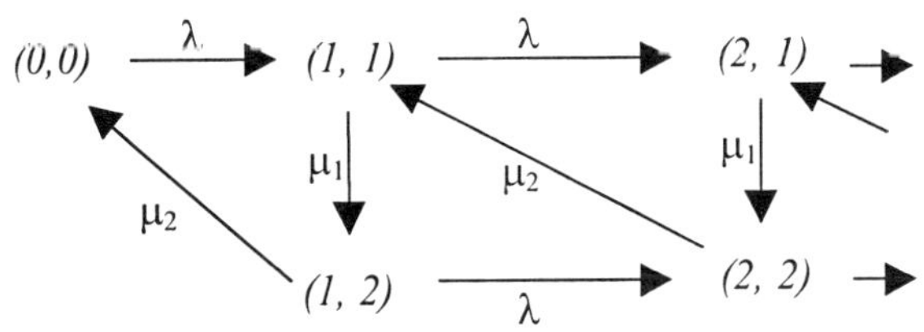

Figure 2.8. State Transition Diagram for Single Sever Queue with Two Stages of Service

The balance equations for this may be written as follows

$$
\begin{aligned}
&\lambda p_{00} = \mu_2 p_{12} \\
&(\lambda + \mu_1) p_{11} = \lambda p_{00} + \mu_2 p_{22} \\
&(\lambda + \mu_2) p_{12} = \mu_1 p_{11} \\
&(\lambda + \mu_1) p_{21} = \lambda p_{11} + \mu_2 p_{32} \\
&(\lambda + \mu_2) p_{22} = \lambda p_{12} + \mu_1 p_{21} \\
&etc.......
\end{aligned}
\tag{2.38}
$$

These may then be solved to get solutions for the state probabilities as a function of p_{00}, the probability of the system being empty, as

$$
\begin{aligned}
p_{12} &= \frac{\lambda}{\mu_2} p_{00} \\
p_{11} &= \frac{\lambda(\lambda + \mu_2)}{\mu_1 \mu_2} p_{00} \\
p_{22} &= \left(\frac{\lambda + \mu_1}{\mu_2} \right) p_{11} - \frac{\lambda}{\mu_2} p_{00} \\
&etc......
\end{aligned}
\tag{2.39}
$$

The complete solution for the various state probabilities may then be found by using the normalisation condition of Eq. (2.4) to first find p_{00} and then the various p_{nj} $n=1,2,......,\infty$, $j=1, 2$. This approach may be easily extended to allow for the following -

(1) Have k Stages of Service Times - For this, we may extend the approach given above to allow *k* stages of service.

(2) Finite Waiting Positions in the Queue - This may be handled by making the arrival rate a function of the number in the system, i.e. make it go to zero when all the waiting positions have been filled.

(3) Multiple Servers - Approximation for this may be done by allowing more than one customer to enter service at a time.

(4) More General Service Distributions - A similar approach may be used to handle more general service time distributions whose Laplace Transforms (of their pdf's) may be represented as a rational function in s. These will be functions of the type $L_B(s)=N(s)/D(s)$ *with simple roots*. For this, consider the system with several stages of service, organised as shown in Figure 2.9

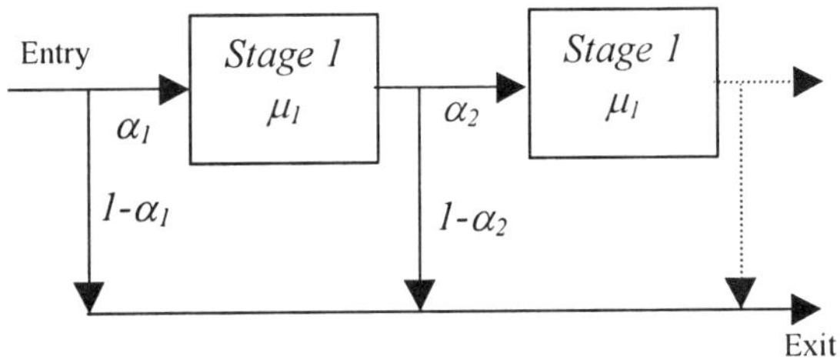

Figure 2.9. Generalised Service in Stages of a Single Server Queue

Consider a general system of the type shown in Figure 2.9 where the individual stages have independent exponentially distributed service times. We assume that the time spent in Stage j is an exponentially distributed random variable with mean $1/\mu_j$ and Laplace Transform $\mu_j/(s+\mu_j)$. The Laplace Transform $L_B(s)$ of the overall service may then be represented as

$$L_B(s) = (1-\alpha_1) + \sum_{j\geq 2} \alpha_1 \ldots\ldots \alpha_{j-1}(1-\alpha_j)\prod_{i=1}^{j} \frac{\mu_i}{s+\mu_i} \tag{2.40}$$

This $L_B(s)$ has simple roots and may be expanded using partial fraction expansion in the following form.

$$L_B(s) = \beta_0 + \sum_i \frac{\beta_i}{s+\mu_i} \tag{2.41}$$

Note that the coefficients β_i $i=0, 1, 2, \ldots\ldots$ may be expressed as functions of the variables μ_i and α_i.

Given a $L_B(s)$ that is a rational function of s with simple roots, we would need to first express it in the form given in Eq. (2.40). This may then be expanded in the form of partial fractions to obtain the form of Eq. (2.41) from which a model using stages may be built as shown in Figure 2.9. This model with service in stages may be analysed in the same way in which we had earlier illustrated the analysis where only two stages were involved. This would involve drawing its state transition diagram with a proper definition of the system state. This state transition diagram may then be used to obtain the corresponding balance equations. These can then be solved to obtain the equilibrium state probabilities of the system. Once the system state probabilities are known, other system parameters may be calculated as desired.

2.10 Queues with Bulk (or Batch) Arrivals

In this case, we assume that the arrivals come in batches (also referred to as bulk arrivals). Effectively, that means that each arrival comes with a random number (say r) of job/customer service requests together. The typical notation for queues of this type is $M^{[X]}/M/1$, $M^{[X]}/M/s/s$, $M^{[X]}/G/1$ etc.. Note that the symbol $M^{[X]}$ used for the arrival process indicates that the batches arrive following a Poisson process but that each batch may have a random number of jobs/customers. In this section, we will look at the $M^{[X]}/M/1$ and the $M^{[X]}/M/s/s$ queues. The $M^{[X]}/G/1$ queue will be considered later in Chapter 3. For the random batch sizes, we assume that β_r is the probability that a batch will have r customers with $r = 1, 2, 3, \ldots\ldots, \infty$, with

$$\textit{Generating Function } \beta(z) = \sum_{r=1}^{\infty} \beta_r z^r \text{ and mean } \bar{\beta} = \sum_{r=1}^{\infty} r\beta_r$$

Note that with the above notation, we have assumed that the batch size is at least one (i.e. the batch will have at least one job). It is possible to use a notation where a *zero-sized batch* is allowed. If this is assumed then the results and derivations given subsequently will need to be appropriately modified. We also assume that λ is the *average arrival rate of batches* to the system. This would, imply that the average arrival rate of a batch of size k would be $\lambda\beta_k$. The *service rate of individual customers* is assumed to be μ where the service times are assumed to be exponential in nature. (The case of generally distributed service times is considered later in Chapter 3.) Note that a batch of size k would then have a *batch service time* that would be the sum of k i.i.d. exponentially distributed random variables - this would be an *Erlang-k* distribution.

2.10.1 The $M^{[X]}/M/1$ Queue

Defining the system state to be the number in the system, the following set of balance equations may be written for this system.

$$\lambda p_0 = \mu p_1 \qquad \textit{for} \quad k = 0$$
$$(\lambda + \mu) p_1 = \mu p_2 + \lambda \beta_1 p_0 \qquad \textit{for} \quad k = 1$$

and

$$(\lambda+\mu)p_k = \mu p_{k+1} + \sum_{i=0}^{k-1} \lambda\beta_{k-i} p_i \qquad for \quad k \geq 1$$

It is easier to use a z-transform based approach here to directly find the *Generating Function P(z)* of the system state. The generating function *P(z)* is defined as

$$P(z) = \sum_{n=0}^{\infty} p_n z^n$$

To obtain this, we would need to multiply the kth balance equation above by z^k and sum from $k=1$ to $k=\infty$. This yields

$$(\lambda+\mu)\sum_{k=1}^{\infty} p_k z^k = \frac{\mu}{z}\sum_{k=1}^{\infty} p_{k+1} z^{k+1} + \sum_{k=1}^{\infty}\sum_{i=0}^{k-1} \lambda p_i \beta_{k-i} z^k$$

Interchanging the order of the double summation and simplifying, we get

$$(\lambda+\mu)[P(z)-p_0] = \frac{\mu}{z}[P(z)-p_0-p_1 z] + \lambda P(z)\beta(z)$$

$$P(z) = \frac{\mu p_0 (1-z)}{\mu(1-z) - \lambda z[1-\beta(z)]}$$

To find p_0, we can use the normalising condition of $P(1)=1$ to get $p_0=(1-\rho)$ with ρ defined to be the average load offered to the system as

$$\rho = \frac{\lambda\bar{\beta}}{\mu}$$

Substituting this value of p_0, the generating function *P(z)* is found to be

$$P(z) = \frac{\mu(1-\rho)(1-z)}{\mu(1-z) - \lambda z[1-\beta(z)]} \tag{2.42}$$

This generating function may then be used to find the individual state probabilities under equilibrium conditions. The moment generating property of the generating function may also be used to directly find the moments of the system state. In particular, one may use Eq. (2.42) to find that the average number N in the system will be

$$N = \frac{\rho(\bar{\beta} + \overline{\beta^2})}{2(1-\rho)} \tag{2.43}$$

where $\overline{\beta^k}$ is the kth moment of the batch size.

2.10.2 The $M^{[X]}$/M/s/s Queue

This corresponds to a s-server queue with no waiting positions and batch arrivals coming from a Poisson process. The service times are exponentially distributed random variables. However, additional details need to be given on the *Batch Acceptance* strategy before this system can be analysed This essentially describes how the system will handle the situation when a batch of size K arrives to a system which has space only for k jobs, with $K>k$. We describe two strategies that one can adopt to handle this situation.

(a) *Partial Batch Acceptance Strategy (PBAS)*
If this strategy is followed, k out of the K jobs in the batch (chosen randomly) will enter the queue, the remaining *(K-k)* jobs will be lost

(b) *Whole Batch Acceptance Strategy (WBAS)*
If this strategy is followed, the entire batch will be refused entry, unless all its jobs can be accepted.

The analysis given below is for an $M^{[X]}$/M/s/s Queue following the PBAS strategy. (A similar procedure may be followed for a queue using a WBAS strategy.) Apart from the notation used earlier for the $M^{[X]}$/M/1 queue, we define ϕ_i as the probability that a batch has i or more jobs in it. (Note that $\phi_1=1$). We can also easily see that

$$\phi_i = \sum_{k=i}^{\infty} \beta_k \qquad \text{for } i=1,\ 2,\ \ldots\ldots.. \tag{2.44}$$

In this case, the Global Balance Equations may be written as follows

$$\lambda \sum_{i=0}^{j-1} p_i \beta_{j-i} + (j+1)\mu p_{j+1} = (\lambda + j\mu) p_j \qquad for \quad j = 1,2,\ldots\ldots,s$$

$$\lambda \sum_{i=0}^{s-1} p_i \phi_{s-i} = s\mu p_s \qquad for \quad j = s \tag{2.45}$$

These balance equations may then be solved to get

$$p_j = \frac{\lambda}{j\mu} \sum_{i=0}^{j-1} p_i \phi_{j-i} \qquad for\ j=1,\ 2,\ \ldots\ldots s \tag{2.46}$$

Using this recursion, we can get the individual state probabilities $\{p_j\}$, as a function of p_0 for any j. For example, one can derive

$$p_1 = \frac{\lambda}{\mu} \phi_1 p_0$$

$$p_2 = \frac{\lambda}{2\mu} [\phi_2 + \frac{\lambda}{\mu} \phi_1^2] p_0$$

Once the state probabilities $\{p_i\}$ are obtained as a function of p_0, we can use the normalisation condition to find first p_0 and then the other state probabilities.

Problems

1. Analyse a queue with a single server where the average arrival rate of customers is $(N-i)\lambda$ per second from a Poisson process, when the system is in state i. Assume that service time required by a customer is exponentially distributed with mean $1/\mu$ seconds. Assume that N is the highest state of the system.

2. The networks lab opens in the morning at 9:00 am. Students arrive at random for the lab following a Poisson process with average rate λ. Each student stays in the lab for a random duration X. Students arriving when there is already a student in the lab, wait outside in the corridor. Find the probability P that the second arriving student will have to wait and also find W, his/her mean waiting time, for the two cases (a) $X=C$ constant and (b) X is exponentially distributed with mean μ^{-1}.

3. Consider an infinite server queue with exponentially distributed service times of mean $1/\mu$.. The arrivals to the queue come from a state dependent Poisson process with the average rate in state i given as α^{i}. Obtain the partial differential equation for $P(z,t)$ that would have to be solved (with proper initial conditions) in order to obtain the time-dependent (transient) state distribution of this system.
Redo the above for the case where the system has finite capacity N, i.e. the state of the system can only be N or less.

4. For a M/M/m/∞ queue, consider a particular customer A, who on arrival finds all m servers busy and n other customers waiting for service. The queue follows a FCFS discipline. Assume that no new customers come to the queue for service after A's arrival. Let $1/\mu$ be the mean service time for customers.
(a) Find the expected length of time customer A spends waiting for service.
(b) Find the expected length of the time interval T measured from A's arrival to the time when the system becomes completely empty.
(c) Define X be the order of completion of service of customer A where $X=k$ if A is the kth customer to complete service after its arrival to the system. Find $P\{X=k\}$ for $k=1,2,\ldots\ldots,(m+n-1)$.
(d) Find the probability that customer A is able to complete its service before the customer who is immediately ahead of him/her in the queue.
(e) Let w be the random amount of time customer A waits for service. Find its *cumulative distribution function* (i.e. c.d.f.).

5. Consider a M/M/2/4 queue at equilibrium. Its state probabilities are observed to be *1/16*, *4/16*, *6/16*, *4/16*, and *1/16* respectively for system states *0, 1, 2, 3,* and *4*. For this queue, determine N and N_q. If the mean arrival rate (from a Poisson process) is observed to be *2* customers per hour, determine the mean delay quantities W and W_q and estimate the mean service time.

6. Consider a M/M/∞/∞ queue where the service rate in state j is given as $\mu_j = j\mu$. The arrival process provides arrivals at an average rate λ. Show that the state probabilities of the system are given by

$$p_0 = e^{-\rho}, \text{ and } p_k = e^{-\rho}\frac{\rho^k}{k!} \quad \text{for } k \geq 1 \qquad \text{with } \rho = \frac{\lambda}{\mu}$$

7. Analyse a M/M/1/∞ queue with the following parameters

$$\lambda_k = \alpha^k \lambda \quad \text{for } k=0,1,2,\ldots\ldots \text{ and } 0 \leq \alpha \leq 1$$
$$\mu_k = \mu \quad \text{for } k=1,2,3\ldots\ldots$$

Obtain the steady state distribution of the queue and the conditions under which such a distribution will exist.

8. Analyse an M/M/1/∞ queue with parameters λ and μ where the customers get impatient and leave if they think they will have to wait too long for service. Specifically, if the arrival find k users already in the system, then it estimates its own waiting time as $w=k/\mu$.. It then either joins the queue with probability $exp(-\alpha w)$ or leaves without service with probability $[1 - exp(-\alpha w)]$. Assume $\alpha>0$.
(a) Find the state distribution of this system and give the conditions under which this will exist.
(b) For $\alpha \rightarrow \infty$, find the state distribution of the system and the average number in the system. Can you give a physical explanation of the system behaviour in this case?

9. The manager of our bank wants to put a new teller system into operation. The proposed system will start operation with only one teller. Whenever the number of customers exceeds K_H, another teller will be brought in to serve the customers. This second teller continues serving until the number of customers falls below K_L where $K_L<K_H$. At this point the second teller leaves and the system continues with one teller. This process continues repetitively. The manager wants our advice on the operation of

this system. Customers arrive at rate λ from a Poisson process and each teller serves with an exponentially distributed service time with mean $1/\mu$..

(a) Give a proper definition of the system state and draw the corresponding state transition diagram with the appropriate transition rates.
(b) Obtain the corresponding state probabilities when the system is in steady state.

10. The Queueing Theory Association is arranging a get-together in a hall where K_1 people can sit and an additional K_2 people can stand. While there is still space available to sit, guests arrive at rate λ_1. When there is only space to stand, guests arrive at rate λ_2. (In both cases, the arrival process is Poisson in nature.) The doors are closed whenever the hall fills up with K_1+K_2 guests. Nobody leaves the get-together once they enter the hall - not even when people start singing badly! Assume that the hall is initially empty when the get-together starts at time $t=0$
(a) Obtain the probability $p_k(t)$ of there being k guests in the hall at time t, $k\geq 0$. (Results may be left in the form of an expression but simplify them to the extent possible.)
(b) From (a) or otherwise, obtain $p_k(t)$ for the case where $K_2=0$.

11. The server in a single-server queue adopts the following approach for serving the customers who arrive for service. Whenever the system becomes empty, the server goes for a coffee-break and comes back to provide service only when the number of customers in the system reaches K. Otherwise, he/she provides service just like a normal M/M/1 queue. Assume λ to be the average customer arrival rate and μ to be the average service rate of a customer being served and let $\rho=\lambda/\mu$ erlangs.
(a) Draw the State Transition Diagram for this system with a proper definition of the system's state.
(b) Now consider specifically the system with $K=2$ and obtain the following in terms of ρ.

- (i) The equilibrium state probability p_k of finding k customers in the system.
- (ii) The mean number in the system (waiting and in-service).
- (iii) The probability that the server is actually working

12. In a M/-/1/∞ queue, the Laplace Transform of the service time distribution is given by

$$L_B(s) = \frac{0.5s(\mu_1 + \mu_2) + \mu_1\mu_2}{(s + \mu_1)(s + \mu_2)}$$

For this queue, find (in terms of the probability p_0 of the system being empty), the probability of there being one user in the system (i.e. p_1) and the probability of there being two users in the system (i.e. p_2). Assume that the arrivals come from a Poisson process of average rate λ.

13. Prof. Calculus has been given the job of registering students for the queueing course in our departmental Conference Room. Prof. Calculus gets a student registered in an exponentially distributed time interval with mean $2/\mu$. He is however known to get very upset if he does not have any additional help when there are *K* or more students in the room (including the one being registered)! When that happens, he calls M. Tintin from our office to help in the registration process. M. Tintin can also register a student in an exponentially distributed time interval with mean $2/\mu$.. When M. Tintin is in the Conference Room, both Prof. Calculus and M. Tintin work in parallel to register students. However, M. Tintin does not quite like this job and manages to return to the office whenever there are no students in the Conference Room. (He gets called again when the number in the room reaches *K*.) Assume that (a) the Conference Room has an infinite number of chairs for waiting students, (b) Prof. Calculus and M. Tintin both work in a FCFS manner and that (c) the students wishing to register arrive from a Poisson process with rate λ.
(a) With a proper definition of system state, draw the state transition diagram for the above queueing situation.
For *K=2,* do the following -
(b) For your states as in (a), obtain the state probabilities of the system.
(c) What will be the mean number of students in the Conference Room?
(d) What is the probability that Prof. Calculus is working on student registration?
(e) What is the probability that M. Tintin is working on student registration?

14. Consider the M/- /1/3 queue, which is limited to having a maximum of 3 users in the system. Assume that the arrivals come from a Poisson process of average rate λ and that the L.T. of the service time distribution is

$$L_B(s) = \frac{2\mu^2}{(s + \mu)(s + 2\mu)}.$$

(a) Draw the *State Transition Diagram* of the system and write its balance equations.
(b) Use (a) to find the probabilities of the individual states in the state transition diagram.
(c) Use (b) to find the overall state probabilities p_0, p_1, p_2 and p_3.
(d) What is the probability that an arrival will have to leave without service?

15. A M/-/1/2 queue has a service time distribution with L.T. given by

$$L_B(s) = \frac{0.5\mu}{s+\mu} + \frac{0.5\mu^2}{(s+\mu)^2}$$

The average arrival rate is λ. Note that the queue is limited to a maximum state of 2. Use the method of stages to solve this queue and obtain the following.
(a) State Transition Diagram (with a proper definition of system states)
(b) Obtain the state probability distribution
(c) What will be the average departure rate from this queue?

16. Consider an examination facility, with an infinitely large waiting room and an examination centre, where queueing theory students are being evaluated. Prof. Calculus and his T.A. for the course sit in the examination centre and students enter the examination centre one at a time. The students (coming from an infinite population of queueing theory students) enter the examination facility at an average rate of λ per time unit. If there is a student already in the examination centre, then the new student waits in the waiting room. The queueing strategy is assumed FCFS. The student entering the examination centre is first examined by the T.A. for an exponentially distributed testing time with mean 0.5 units. Prof. Calculus randomly picks half of the students examined by the T.A. and grills them further for an exponentially distributed time of 1.0 units. (The others are lucky and can leave immediately without being examined by Prof. Calculus.) Assuming that the probability p_0 of the system being empty is known, find the following.
(a) The probability that there is one student in the examination facility
(b) The probability that there are two students in the examination facility

17. Consider a $M^{[X]}/M/1$ queue where the batch arrival process is Poisson with rate λ and the batch size distribution is given by $\beta_i = P\{batch\ size = i\}$ for $i \geq 0$ with $\beta(z)$ as its z-transform. Assume that the mean service time for a job is $1/\mu$.

(a) Draw the State Transition Diagram of the system and write its balance equations.
(b) Find the generating function *P(z) of* the number in the system at equilibrium as a function of λ, μ and *G(z)* with

$$\rho = \frac{\lambda\beta'(1)}{\mu}$$

defined as the offered load.
(c) For the case where $\beta_i = (1-\alpha)\alpha^i$ for $0<\alpha<1$ and $i = 0, 1, \ldots, \infty$ find *P(z)* and use this to get the state probability distribution p_k $k=0, 1, \ldots, \infty$.
(d) Find the mean number of jobs in the system. Use this and Little's result to find the mean time spent in the system by a particular job.
(e) Assuming FCFS operation, derive the mean time spent waiting for service to begin to a particular batch. (How would you find this?) Can I use this to find the mean time spent in system by a particular job?

18. Consider *a* $M^{[X]}/M/1/4$ queue with batch arrival rate λ and service rate of individual jobs as μ. The Generating Function of the batch sizes is given to be $(\beta_0 + \beta_1 z + \beta_2 z^2)$. Analyse this system for the cases where *(i) PBAS* and *(ii) WBAS* strategies are being used. For each case, obtain the following.
(a) State probabilities of the system
(b) The mean number in the system
(c) The probability that a batch is refused entry in the WBAS case and the probability that a job is refused entry in the PBAS case.

19. Consider a $M^{[X]}/-/1/3$ queue where the batch arrival rate is λ and the generating function of the batch sizes is given by $(0.25z+0.25z+0.5z^2)$. Note that the queue is limited to a maximum state of *3* and follows the *WBAS* strategy.
(a) If the L.T. of the service time distribution is

$$L_B(s) = \frac{0.5\mu}{s+\mu} + \frac{0.5\mu^2}{(s+\mu)^2}$$

draw the state transition diagram of the queue with an appropriately defined system state.
(b) For a service time distribution with L.T. $\mu/(s+\mu)$, do the following -

i. Draw the state transition diagram.
ii. Find the state probabilities of the queue.
iii. What is the probability that a batch is refused entry into the queue?

Chapter 3

Analysis of the M/G/1 Queue in Equilibrium

Performance Analysis Using Residual Life and Imbedded Markov Chain Approaches

In the previous chapter, we were primarily concerned with queues where the service times were exponentially distributed. Queues with non-exponential service times were not really considered, except in a limited fashion in Section 2.9 while describing the Method of Stages. In this chapter, we consider the detailed analysis of single server queues with infinite buffers where the service times can have any general distribution. The arrival process is still assumed to be Poisson in nature.

The exponential distribution is particularly easy to handle in analytical modelling because of its memory-less property. Consider a situation where the service time is a random variable X with mean μ^{-1}. If a job in service is examined at any time while its service is continuing, the distribution of the remaining service time will still be exponentially distributed with the same mean value μ^{-1} as before. This essential feature simplifies the analysis of a queue by allowing us to define the system state at any time instant t by using just a single variable, i.e. the number in the system at time t. This simple definition is no longer possible in the general case where the system state has a non-exponential distribution. If that is the case, the system state at an arbitrary time instant t must be defined as an *n-tuple*. This would consist of both the number in the system at time t as well as the residual service times for each customer currently in service. This additional complexity makes the non-exponential service time distribution case much harder to analyse.

Unfortunately, though the assumption of exponential distribution is convenient for analytical purposes, it may not be justified in all systems. For situations where the service time can be expressed, at least approximately, as a combination of exponentially distributed components, the Method of Stages of Section 2.9 may be used. The approach given in that section may be suitably modified to allow, at least approximately, for more than one server or limited waiting spaces in the queue. Exact analytical modelling is however possible for generally distributed service times in the case of single server queues with infinite buffers, i.e. M/G/1 queues. This may be done in a number of ways. We present two such methods in this chapter - the *Residual Life Approach* and a method using an *Imbedded Markov Chain*. The Residual Life Approach is simpler but can only give mean results on the mean queueing parameters W, W_q, N and N_q. The method using an imbedded Markov chain is more powerful and may be used to obtain distributions of the various parameters, as well as their mean values. This latter approach analyses a M/G/1 queue by identifying a special sequence of time instants. These are such that the Markovian property of being memory-less will be satisfied for the system state (defined as the number in the system) between these time instants. In this case, the intelligent choice of these time instants, enforcing the Markovian property, simplifies the system state definition and the consequent analysis.

Unless specifically indicated otherwise, we consider the M/G/1queue to be FCFS in nature. (This does not make any difference to the mean values of the system parameters but would affect higher moments and distributions.) The arrival process is considered Poisson with average arrival rate λ. The inter-arrival times will be exponentially distributed with mean $1/\lambda$ and variance $1/\lambda^2$. The interarrival times will have the cumulative distribution function $A(t)$ and probability density function $a(t)$ respectively given by

$$\begin{aligned} A(t) &= 1 - e^{-\lambda t} && \text{for } t \geq 0 \\ &= 0 && \text{for } t < 0 \\ a(t) &= \lambda e^{-\lambda t} && \text{for } t \geq 0 \\ &= 0 && \text{for } t < 0 \end{aligned}$$

The Laplace Transform (L.T.) $L_A(s)$ of the probability density function will then be -

$$L_A(s) = \frac{\lambda}{s + \lambda}$$

The service time X is considered to be generally distributed with mean $E\{X\} = \overline{X}$. For $X=t$ $(t \geq 0)$, the distribution is given in terms of its cumulative distribution function $B(t)$, probability density function $b(t)$ or its L.T. $L_B(s)$. We assume that the distribution is known. The analysis given here is under equilibrium conditions when $\rho = \lambda\overline{X} < 1$ holds.

3.1 The Residual Life Approach for Analysing the M/G/1 Queue

For convenience, we will assume that the queue serves customers using the FCFS service discipline. As mentioned earlier, this service discipline will not make a difference to the mean performance parameters as long as the server does not stay idle in a situation where there are one or more customers in the queue. This implies that the mean performance results will remain the same even if we use LCFS or SIRO service disciplines. A queue of this kind is shown in Figure 3.1.

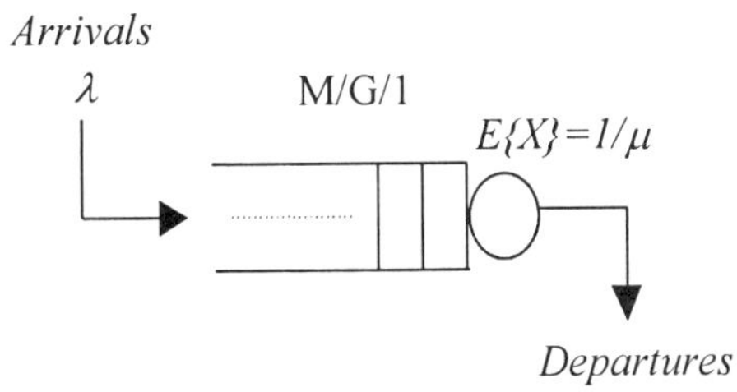

Figure 3.1. A M/G/1 Queue

For the queue shown, consider a particular arrival of interest that enters the queue. If it arrives to a non-empty queue, it will find -

(a) one or more customers waiting in the queue, and

(b) a customer currently in service who would have r seconds of *residual service time* left.

Note that the residual service would be the additional time required to finish service to the customer currently being served. If the arrival finds the queue empty, then the residual service time it will see will also be zero and it will enter service immediately.

Let R=E$\{r\}$ be the mean residual service time. This will be the mean of the random variable r as observed by arrivals to the queue. Note that for this queue, PASTA property will hold, as the arrival process is Poisson in nature. Therefore, using Little's result, we can write

$$W_q = N_q E\{X\} + R = \lambda W_q E\{X\} + R$$

Defining $\rho=\lambda E\{X\}$ as before, we require $\rho<1$ for the queue to be stable. In that case, we can simplify the above to get

$$W_q = \frac{R}{(1-\rho)}$$

To use this to find W_q, we would need the mean residual service time R. As shown in [BeG92], this may be conveniently found using a graphical approach. To apply this approach, we first observe that the value of R as seen by the arrival will also be the mean residual service time as calculated by observing the queue in steady state and taking time averages. This would be true because PASTA would hold for this queue, as its arrival process is Poisson in nature. We define $r(\tau)$ as the value of the residual service time that will be seen if the system is examined at time τ. The function $r(\tau)$ is plotted below as a function of τ. The *saw-toothed* nature of the plot arises because

(a) $r(\tau)$ jumps by an amount equal to the required service time whenever a new customer starts service, and

(b) when $r(\tau)$ is non-zero, it decreases at a unit rate until it reaches zero (when it reaches zero the customer being served completes service).

These properties are illustrated in Figure 3.2.

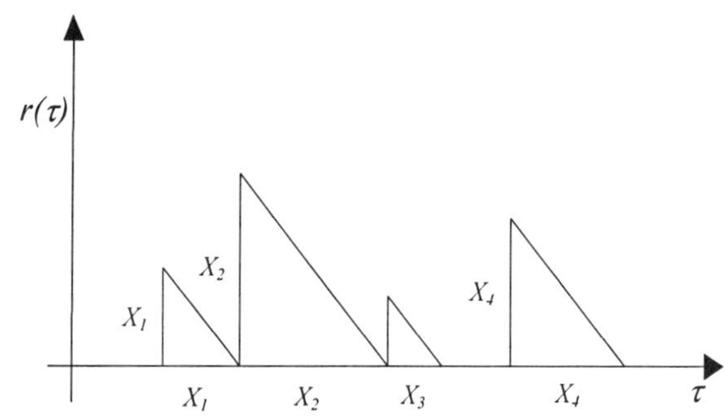

Figure 3.2. Residual Service Time $r(\tau)$ as a Function of τ

Let $M(t)$ be the number of departures from the queue in the time interval $(0, t)$. We can then find the time average of $r(\tau)$ for $0\leq \tau\leq t$ as follows

$$\text{Time average of } r(\tau) \text{ in } (0,t) = \frac{1}{t}\int_0^t r(\tau)d\tau \cong \frac{1}{t}\sum_{i=1}^{M(t)}\frac{1}{2}X_i^2$$

Note that because of the ergodicity implied by the PASTA property, the limit of this average as $t \to \infty$ will then be equal to R. The approximation above holds by ignoring the errors, if any, in the last residual service time in the interval $(0, t)$. Note that this approximation becomes better as $t \to \infty$. We can then write

$$\text{Time average of } r(\tau) \text{ in } (0, t) = \frac{1}{t}\sum_{i=1}^{M(t)} \frac{1}{2} X_i^2 = \frac{1}{2} \frac{M(t)}{t} \frac{1}{M(t)} \sum_{i=1}^{M(t)} \frac{1}{2} X_i^2$$

$$\text{For } t \to \infty \text{, we get } \frac{M(t)}{t} \to \lambda \text{, and } \frac{1}{M(t)} \sum_{i=1}^{M(t)} \frac{1}{2} X_i^2 \to \overline{X^2}$$

Therefore, by PASTA,

$$R = \frac{1}{2} \lambda \overline{X^2} \tag{3.1}$$

$$W_q = \frac{\lambda \overline{X^2}}{2(1-\rho)} \quad \text{(Pollaczek-Khinchine or P-K Formula)} \tag{3.2}$$

Incidentally, the graphical approach given above would be found to be a useful trick to use for analysing many queueing situations. We can also make an interesting observation on R by decomposing it as follows, using the fact that the residual service time observed by an arrival to an empty queue will be zero. For this note that as obtained earlier, the load on the queue will be given by $\rho = \lambda/\mu$ and that ρ and $(1-\rho)$ will be the probability that arrival finds the queue non empty and the probability that arrival finds the queue empty, respectively.

This is easy to argue regardless of the actual nature of the arrival and service processes. Consider a long interval of time T, with $T \to \infty$. The average number of arrivals in that interval would be λT and each arrival will require a mean service time of $E\{X\}$. Therefore, if the queue is a stable one, then out of the total time T, a time of $(\lambda T)E\{X\}$ will be observed to be the average time the server is busy. This implies that the P{server is busy}$=\rho=\lambda E\{X\}$ and P{server not busy}$=(1-\rho)$.

Decomposing R into the two cases mentioned above, i.e. where arrival finds an empty system and where arrival finds a non-empty system, then gives

R = $(1\text{-}\rho)$E{r | system found empty on arrival} + ρE{r | system found not empty on arrival}

Since E{r| system found empty on arrival} = 0, then using R obtained in Eq. (3.1), we get that

$$\text{E}\{r \mid \text{system found not empty on arrival}\} = \frac{R}{\rho} = \frac{\overline{X^2}}{2\overline{X}} = \frac{1}{2}\left(\overline{X} + \frac{\sigma_X^2}{\overline{X}}\right)$$

This result is actually somewhat counter-intuitive. Since the arrival coming to a non-empty queue essentially samples an ongoing service, lay reasoning may lead us to expect that the mean value of the residual service time observed in that case should be $\frac{1}{2}\overline{X}$ rather than the $\frac{1}{2}\left(\overline{X} + \frac{\sigma_X^2}{\overline{X}}\right)$.

This is also popularly referred to (in the context of other phenomenon) as the *Paradox of Residual Life*. This is not really a paradox - as will be explained subsequently. Instead, this follows from the fact that the arrival (to a non-empty queue) is more likely to sample a longer service time than a shorter one.

3.1.1 The Paradox of Residual Life

The paradox of residual life is commonly illustrated in another way by observing arrivals from a process, whose inter-arrival times have some general distribution, i.e. it is not necessarily a Poisson process with the inter-arrival times exponentially distributed. Consider the process of arrivals as illustrated in Figure 3.3. As shown in the figure, let the time instant t_k be the instant of the kth arrival from the arrival process. Let the random variable X represent the time between successive arrival instants, i.e. the inter-arrival time of the process. Its mean and second moment will then be given by

$$E\{t_k - t_{k-1}\} = \overline{X}$$

$$E\{(t_k - \tau_{k-1})^2\} = \overline{X^2}$$

where the inter-arrival time X is considered to be a *lifetime*.

We examine this process at an arbitrary time t, as shown. Assume that t happens to fall in the interval (t_k, t_{k+1}) and that the random variable Y

corresponding to the *Residual Lifetime* is given by $Y = (t_{k+1} - t)$. Drawing an analogy with the results obtained earlier for the residual service time when the arrival comes into a non-empty queue, we can see that

$$\overline{Y} = \frac{1}{2}\overline{X} + \frac{1}{2}\frac{\sigma_X^2}{\overline{X}} \tag{3.3}$$

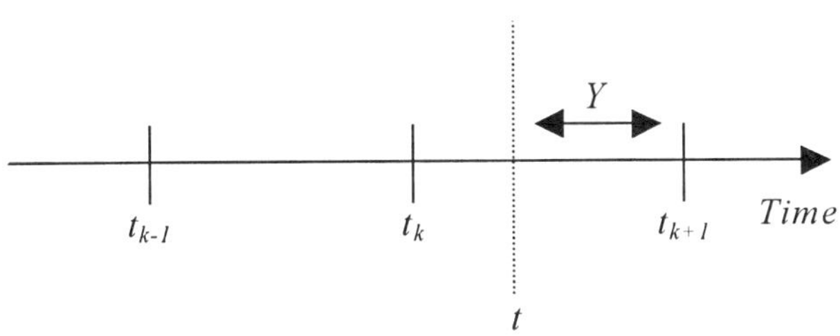

Figure 3.3. Arrival Process Illustrating the Paradox of Residual Life

The *Paradox of Residual Life* may then be stated as follows. "Since *t* can be chosen anywhere, we would expect $\overline{Y} = 0.5\overline{X}$. However, this is not really the case as we would find that (as obtained above!) actually $\overline{Y} \geq 0.5\overline{X}$ and that $\overline{Y}$ increases with increasing σ_X (i.e. with increasing randomness)".

The subject of *Renewal Theory* would look at this residual life from a somewhat different (but essentially similar) perspective. This would consider a system where components fail at time t_i, $i=1, 2,$ ...... and are immediately replaced. The system is observed at some random time *t* and the observer looks at the residual life *Y* of the current component. Assume that we are given that $X=(t_{k+1} - t_k)$ is the lifetime of the kth component with its distribution (cumulative distribution function $F_X(x)$ and probability density function $f_X(x)$) defined as follows

$$F_X(x) = P\{(t_{k+1} - t_k) \leq x\}$$
$$f_X(x) = \frac{dF_X(x)}{dx} \qquad \forall k$$

We would then like to find the distribution $f_Y(y)$ of the random variable *Y* (the residual lifetime) and its mean. Consider X^* to be the lifetime (random variable) of the component being examined at time t (i.e. the selected component). We would first need to find the probability density function $f_{X^*}(x)$ of this random variable. For this consider the probability of the event that $\{t$ falls in a lifetime of length $x\}$. (Note that time t is the instant at which the system is observed.) We can observe the following.

(a) P$\{t$ falls in a life time of length $x\}$ would be proportional to the length x of the lifetime, as t is more likely to fall in a longer lifetime than in a shorter one.
(b) P$\{t$ falls in a life time of length $x\}$ would also be proportional to $f_X(x)dx$, i.e. the probability of occurrence of a lifetime of length x.
(c) P$\{t$ falls in a life time of length $x\}$=P$\{$a lifetime of length x is selected$\}$ $=f_{X^*}(x)$

Therefore, we get that,

$$\text{P}\{t \text{ falls in a life time of length } x\} = f_{X^*}(x)dx = Kxf_X(x)dx$$

where K is a normalising constant. Integrating $f_{X^*}(x)dx=Kxf_X(x)dx$ over the range $(0, \infty)$ for x, we get

$$K = 1/E\{X\} = 1/\overline{X}$$

The probability density function of the lifetime of the selected component will then be given as

$$f_{X^*}(x) = \frac{xf_X(x)}{\overline{X}} \tag{3.4}$$

in terms of the probability density function and mean of the component lifetime (overall, i.e. unselected) X.

Now consider the residual lifetime Y. Note that the residual lifetime arises by the selection of t in a uniformly distributed fashion over the lifetime of the *selected component*. Therefore,

$$P\{Y \le y \mid X^* = x\} = \frac{y}{x} \quad \text{for } 0 \le y \le x \tag{3.5}$$

or

$$P\{y \le Y \le y+dy, x \le X^* \le x+dx\} = \frac{dy}{x}\frac{xf_X(x)}{\overline{X}}dx$$
$$= \frac{f_X(x)}{\overline{X}}dxdy \qquad for \quad 0 \le y \le x \tag{3.6}$$

Integrating Eq. (3.6) with respect to x over $y \le x \le \infty$, we get the required probability density function $f_Y(y)$ of the residual lifetime Y to be

$$f_Y(y)dy = P\{y \le Y \le y+dy\} = \frac{1}{\overline{X}}[1-F_X(y)]dy \tag{3.7}$$

The Laplace Transform $L_Y(s)$ of the probability density function of the residual lifetime Y may then be computed from Eq. (3.7) to be

$$L_Y(s) = L.T.\left(\frac{1}{\overline{X}} - \frac{1}{\overline{X}}\int_0^y f_X(y)dy\right) = \frac{1-L_X(s)}{s\overline{X}} \tag{3.8}$$

where $L_X(s)$ is the Laplace Transform of the probability density function of X. Differentiating this, we get

$$L'_Y(s) = -\frac{1-L_X(s)+sL'_X(s)}{s^2\overline{X}} \tag{3.9}$$

Using Eq. (3.9) and the moment generating property of the *Laplace Transform*, the *Mean Residual Lifetime* $E\{Y\} = \overline{Y}$ may be found using $\overline{Y} = -L'_Y(s)\big|_{s=0}$. This actually becomes indeterminate when evaluated at $s=0$, so *L'Hospital's Rule* has to be used to evaluate this. Using this and the fact that $\overline{X} = -L'_X(s)\big|_{s=0}$ and $\overline{X^2} = L''_X(s)\big|_{s=0}$, we get

$$\overline{Y} = \lim_{s\to 0}\left(\frac{-L'_X(s)+L'_X(s)+sL''_X(s)}{2s\overline{X}}\right)$$
$$= \frac{\overline{X^2}}{2\overline{X}} = \frac{1}{2}\left(\overline{X}+\frac{\sigma_X^2}{\overline{X}}\right) = \frac{1}{2}\overline{X}+\frac{\sigma_X^2}{2\overline{X}} \tag{3.10}$$

Note that this is what we had also obtained earlier with the graphical approach.

3.2 The Imbedded Markov Chain Approach for Analysing the M/G/1 Queue

For M/M/-/- type of queues, we were able to represent the system state at time t as the number in the system at time t and were able to use this to form an imbedded Markov Chain at any arbitrary set of time points. For the M/G/1 queue, this cannot be done, as the service time is no longer memory-less. If we did want to persist with this approach then we need to expand the system descriptor appropriately. One way to do this would be to describe the system as say (n, α). Here n would be the number in the system at time t and α would be the time left to finish service to the customer currently in service at time t. Note that we did not have to do this for a system with exponentially distributed service times because of the memory-less property of that distribution.

However, even for the M/G/1 queue, a single system state descriptor will work if we choose our time points carefully so that the Markov Property is satisfied between these time points. One such set of time points satisfying the Markov property are the *time instants just after the departure of a customer following service*. Consider the imbedded Markov Chain of system states (denoting the number in system) at the time instants t_i $i=1, 2, 3, \ldots\ldots \infty$ when the ith customer departs from the system. At a time instant t_i, we define the system state n_i to be the number of customers left behind when the ith customer departs (i.e. immediately after the departure of the ith customer).

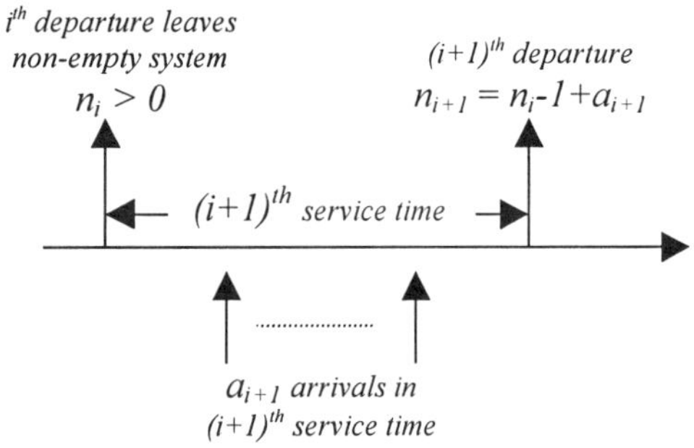

Figure 3.4a. Departure Leaves System Non-Empty

Note that the system state is actually what the departing customer would see looking back into the system after it leaves (the system) - that is it does not count itself in the number indicating the system state. Let a_{i+1} be the number of arrivals in the $(i+1)^{th}$ service time. Note that the $(i+1)^{th}$ service time will end with the departure of the $(i+1)^{th}$ customer. Two situations arise here, as shown in Figure 3.4a and 3.4b, and must be considered separately.

In the first case, shown in Figure 3.4a, the ith departure leaves behind a non-empty queue. Therefore the time interval between the $(i+1)^{th}$ and the ith departure is just one (random) service time with a_{i+1} arrivals in that interval as per the definition given above. Note that the state at the $(i+1)^{th}$ departure instant can be related to the state at the ith departure instant thereby forming a Markov Chain. Specifically, the state at the $(i+1)^{th}$ departure instant is $n_i - 1+a_{i+1}$.

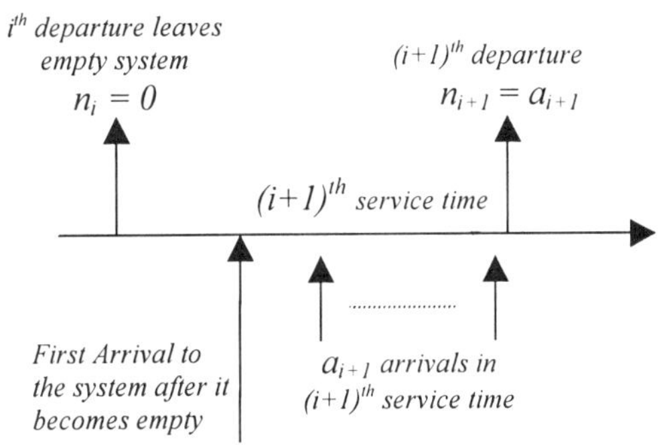

Figure 3.4b. Departure Leaves System Empty

For the case where the ith departure leaves the queue empty, service does not start again (nor are any departures possible) until the $(i+1)^{th}$ arrival. This is the customer whose departure will be tagged as the $(i+1)^{th}$ departure as shown in Figure 3.4b. The number left behind in the queue after this customer departs will merely be the number arriving during the service time of the customer. Therefore, the system state n_i at the $(i+1)^{th}$ departure is merely a_{i+1}.

From the above, it can be seen that

$$\begin{aligned} n_{i+1} &= a_{i+1} && for \quad n_i = 0 \\ &= n_i - 1 + a_{i+1} && for \quad n_i = 1,2,3......... \end{aligned} \tag{3.11}$$

Sometimes, it may also be convenient to rewrite Eq. (3.11) as

$$n_{i+1} = n_i - U(n_i) + a_{i+1} \qquad for \quad n_i = 0,1,2,3,........ \tag{3.12}$$

where $U(n_i)$ is the unit step function.

It should be noted that for the queue to be stable we should once again have $\rho=\lambda/\mu<1$. This can be justified either based on the earlier arguments for stability of a single server queue or by requiring (equivalently) that *the average number of customers arriving in a service time should be less than 1* for the system to be stable. The results obtained below hold when this stability condition is met.

For the M/G/1 queue being considered, Eq. (3.12) may now be used to obtain the probabilities of the system state as observed by a departing customer. (We do this subsequently.) At first sight, these probabilities do not seem to be too useful as they hold only for the departure instants. It is not immediately obvious how these results may be used to find ergodic (i.e. time-averaged, steady state) distributions and moments. We need the following to help us put these results to more practical use.

1. *Kleinrock's Result*:

This states that for systems where the system state can change at most by +1 or -1, the system distribution as seen by an arriving customer will be the same as that seen by a departing customer.

We can use this result to claim that the distributions or moments computed from the point of view of the departing customer will therefore be the same as the state distributions and moments as seen by an arriving customer.

2. *PASTA*

Since the arrival process is Poisson (i.e. the queue is M/G/1), PASTA will hold. Therefore the state distribution and moments as seen by the arriving customer will also be the same as the time averages observed under equilibrium conditions.

Using both Kleinrock's Result and the PASTA property, we can therefore claim that the state distribution and moments calculated for the departure instants will also be the time averaged (ergodic) results that will be observed for the system at equilibrium. This is what makes the results obtained using Eq. (3.12) useful for studying the equilibrium performance of a M/G/1 queue. Some typical results are derived next based on the system description given by Eq. (3.12).

3.2.1 Probability of the System Being Empty

For obtaining this, we take the expectations of both sides of Eq. (3.12) to get that $E\{U(n_i)\}=E\{a_{i+1}\}$. We can then see that since

$$E\{U(n_i)\} = 1 - p_0 \quad \text{and} \quad E\{a_{i+1}\} = \int_0^\infty (\lambda t) b(t) dt = \lambda \overline{X} = \rho$$

we get our expected result that $p_0 = 1 - \rho$

3.2.2 Generating Function of the Number in the System (the P-K transform equation)

We define the generating functions $P_i(z)$ and $P_{i+1}(z)$ as follows where $P_j(z)$ is the generating function of the state as seen by the jth departing customer.

$$P_i(z) = E\{z^{n_i}\} = \sum_{k=0}^{\infty} z^k P\{n_i = k\}$$

$$P_{i+1}(z) = E\{z^{n_{i+1}}\} = \sum_{k=0}^{\infty} z^k P\{n_{i+1} = k\}$$

Since future arrivals would not depend on the current state, we would have n_i to be independent of a_i (i.e. $n_i \perp a_i$). We would therefore get that

$$P_{i+1}(z) = E\{z^{n_i - U(n_i)}\} E\{z^{a_{i+1}}\}$$

This can be simplified by observing that -

(1) Since the system is being examined in steady state (equilibrium) conditions, the dependence on "i" may be dropped.

(2) If we define $A(z)=E\{z^a\}$ as the generating function of the number arriving in a service time, then we would get that

$$A(z) = \int_0^\infty E\{z^a \mid service\ time = t\} b(t) dt = \int_0^\infty \sum_{n=0}^{\infty} z^n \frac{(\lambda t)^n}{n!} e^{-\lambda t} b(t) dt$$

$$= \int_0^\infty e^{-\lambda t(1-z)} b(t) dt$$

leading to

$$A(z) = L_B(\lambda - \lambda z) \tag{3.13}$$

It is interesting to note that the result of Eq. (3.13) could have been written down directly as *A(z)* is the generating function of the number arriving in a service time, where the arrival process is Poisson with rate λ. This may be done following the approach taken to derive Eq. (2.36) in Chapter 2 and the associated comments given there.

Using Eq. (3.13) and the moment generating properties of both generating functions and Laplace Transforms, we get that

$$A'(z) = -\lambda L'_B(\lambda - \lambda z) \quad A'(1) = -\lambda L'_B(0) = \lambda \overline{X} = \rho$$
$$A''(z) = \lambda^2 L''_B(\lambda - \lambda z) \quad A''(1) = \lambda^2 L''_B(0) = \lambda^2 \overline{X^2}$$

Using Eq. (3.12), we can then obtain

$$\begin{aligned} P(z) &= A(z)E\left\{z^{n-U(n)}\right\} = A(z)\sum_{k=0}^{\infty} z^{k-U(k)} P\{n=k\} \\ &= A(z)\left[z^0 p_0 + \sum_{k=1}^{\infty} z^{k-1} p_k\right] = A(z)\left[p_0 + \frac{1}{z}\sum_{k=0}^{\infty} z^k p_k - \frac{1}{z} p_0\right] \\ &= A(z)\left[\frac{1}{z} P(z) - \frac{1}{z} p_0 (1-z)\right] \end{aligned}$$

Using $p_0 = 1-\rho$, and rearranging terms, this is simplified to give the Generating Function *P(z)* of the number in the system as seen by a departing customer (at equilibrium) to be

$$P(z) = \frac{(1-\rho)(1-z)A(z)}{A(z)-z}$$

where *A(z)* may be found using Eq. (3.13). Substituting for *A(z)* and simplifying, we get the final expression for the generating function

$$P(z) = \frac{(1-\rho)(1-z)L_B(\lambda - \lambda z)}{L_B(\lambda - \lambda z) - z} \qquad \textit{P-K Transform Equation} \tag{3.14}$$

Note that the generating function *P(z)* may be inverted (i.e. its inverse z-transform may be computed) to find the corresponding state distribution. This distribution will not only hold for the system observed at the departure instants, but will also hold for the system observed at the arrival instants. It will also be the ergodic state distribution. Note that even if *P(z)* may not be exactly inverted in a closed-form expression, we can obtain the desired distribution by expanding it in a Taylor Series expansion as

$$P(z) = \sum_{i=0}^{\infty} \alpha_i z^i$$

Once this is done by suitably manipulating the R.H.S of Eq. (3.14), we can identify α_i as the equilibrium probability p_i of the system being in state *i* (i.e P{*i* in system}. Various moments of the system state can then be found using these probabilities. It will, of course, be far simpler to use the generating function *P(z)* directly to find the required moments. This is illustrated next.

3.2.3 Computing Moments of the System Parameters for the M/G/1 Queue

i) Note that *P(1)=A(1)=1*, and that as obtained previously

$$A'(1) = \lambda \overline{X} = \rho \quad \text{and} \quad A''(1) = \lambda^2 \overline{X^2}$$

$$\text{ii) } P(1) = \lim_{z \to 1} P(z) = \lim_{z \to 1} \frac{(1-\rho)[(1-z)A'(z) - A(z)]}{A'(z) - 1} = -\frac{(1-\rho)}{\rho - 1} = 1$$

Note that evaluation of the above limit requires the use of L'Hospital's Rule. The result *P(1)=1* is expected as the total sum of all the state probabilities must be unity.

iii) From the P-K Transform Equation, we can obtain the mean and higher moments as well

$$P(z)[A(z) - z] = (1-\rho)(1-z)A(z)$$
$$P'(z)[A(z) - z] + P(z)[A'(z) - 1] = (1-\rho)(1-z)A'(z) - (1-\rho)A(z)$$
$$P''(z)[A(z) - z] + 2P'(z)[A'(z) - 1] + P(z)A''(z) = (1-\rho)(1-z)A''(z) - (1-\rho)A$$
$$\ldots\ldots\ldots$$

Putting $z=1$ and using earlier results, we get

$$-2P'(1)(1-\rho) + \lambda^2 \overline{X^2} = (1-\rho)(-2\rho)$$
$$N = P'(1) = \rho + \frac{\lambda^2 \overline{X^2}}{2(1-\rho)} \qquad \textit{Mean number in system}$$

Other higher moments may be similarly found by further differentiation of $P(z)$. Note that this will then also need evaluation of higher differentials of $A(z)$ for $z=1$. The other mean system performance parameters, i.e. W, W_q and N_q, may now be found from N and are given below.

$$\text{Mean Time Spent in System} = W = \overline{X} + \frac{\lambda \overline{X^2}}{2(1-\rho)} \qquad \text{(using Little's Result)}$$

$$\text{Mean Time Spent in Queue} = W_q = W - \overline{X} = \frac{\lambda \overline{X^2}}{2(1-\rho)}$$

$$\text{Mean Number Waiting in Queue } N_q = \frac{\overline{X^2}}{2(1-\rho)} \qquad \text{(using Little's Result)}$$

It should be noted that the expression given above for W_q is the same as the mean result (P-K Formula) obtained earlier in Eq. (3.2) using the residual life approach.

3.3 Distributions of Time Spent in System and the Waiting Time Prior to Service in a FCFS M/G/1 Queue

Given the distribution for the number in the system as seen by a departing customer, it is fairly straightforward to derive the distribution of time spent waiting in queue and total time spent in system by a customer.

We outline the approach next as this approach is very commonly used in analysing the delays of a queueing-type system. The distribution for a LCFS queue is much more difficult to obtain and will be postponed for Section 3.5

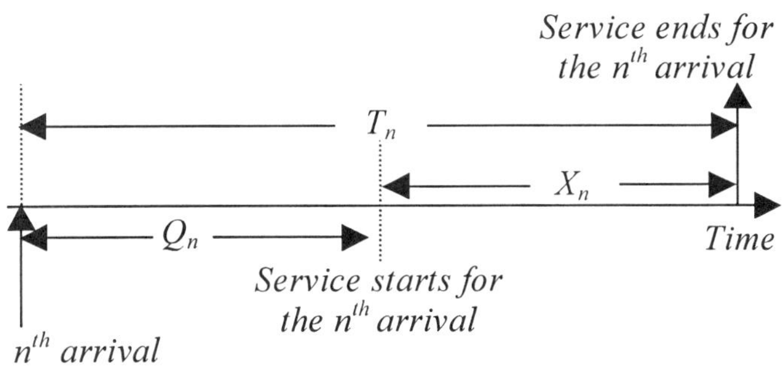

Figure 3.5. Time Instants of Arrival and Departure for a Customer in a FCFS M/G/1 Queue

For a FCFS M/G/1 queue, consider the nth arrival to the queue as shown in Figure 3.5. As shown in the figure, this arrival waits in the queue for a time interval of length Q_n before its service can start. (Note that $Q_n=0$ if the arrival enters an empty queue.) Once service starts, the customer engages the server for a time interval X_n corresponding to its service time. The total time T_n spent in the system by the nth arrival is then Q_n+X_n.

As before, we use $b(t)$ to denote the probability density function of the service time X_n with its L.T. given by $L_B(s)$. Similarly, let $f_Q(t)$ be the probability density function of the queueing delay Q_n with L.T. $L_Q(s)$ and let $f_T(t)$ be the probability density function of the total time T_n spent in system by the nth arrival with L.T. $L_T(s)$. It is important to note here that the probability density function's of these random variables and their L.T.s have been written without explicitly mentioning the variable n (for the nth arrival) since the system is assumed to have reached equilibrium conditions.

Consider the nth arrival as shown in Figure 3.5. Since the queue is assumed to be FCFS in nature, the number of customers that the nth user will see left behind in the queue when it departs will be the number of arrivals that occur while it is in the system. The generating function for this random number (at equilibrium) has already been obtained as $P(z)$ in Eq. (3.14) as the generating function of the number in the system as seen by a departing customer.

We can also argue that the generating function $P(z)$ for the number in the system when this n^{th} customer departs can also be obtained in another way.

Using the approach used to get Eq. (2.36) and Eq. (3.13), we can argue that this will be given as $L_T(\lambda\text{-}\lambda z)$ where $L_T(s)$ is the L.T. of the probability density function of the total time T spent in the system by an arriving customer. Therefore, we get that

$$L_T(\lambda - \lambda z) = \frac{(1-\rho)(1-z)L_B(\lambda - \lambda z)}{L_B(\lambda - \lambda z) - z}$$

Substituting $s=(\lambda - \lambda z)$, we then get

$$L_T(s) = \frac{s(1-\rho)L_B(s)}{s - \lambda + \lambda L_B(s)} \tag{3.15}$$

The L.T. $L_T(s)$ may then be inverted to obtain the probability density function $f_T(t)$ of the total time spent by an arriving customer in a FCFS M/G/1 queue. It is also possible to directly use $L_T(s)$ to obtain the desired moments of the random variable T, the total time spent in system by an arriving customer.

Knowing $L_T(s)$, and using the fact that $Q+X=T$, where Q and X are independent of each other (i.e. $Q \perp X$), we get

$$L_Q(s) = \frac{L_T(s)}{L_B(s)} = \frac{s(1-\rho)}{s - \lambda + \lambda L_B(s)} \tag{3.16}$$

as the L.T. of the probability density function $f_Q(s)$ of the queueing delay Q. Inverting this Laplace Transform will give the associated probability density function $f_Q(t)$ of the queueing delay. Even if this inversion is not explicitly done, the moments of Q may be found directly from $L_Q(s)$ itself by differentiating it and evaluating this at $s=0$ using the moment generating properties of Laplace Transforms.

3.4 Busy Period Analysis of a M/G/1 Queue

We can observe the server of this single-server queue to see the time intervals during which it is is busy or idle. If we observe the queue in this fashion, we will notice that the time-axis may be considered to be divided into *cycles* of random length. Each *cycle* consists of a sequence of an *idle period* (when the queue is empty and the server is idle) followed by a *busy period* (during which the server is busy and the queue is non-empty).

The busy period may also be viewed on the basis of a variable, *Unfinished Work U(t)*, that is sometimes convenient to consider in a queue. We define $U(t)$ at time t as the amount of time that it would take to *empty the system of all those customers who were present in the system at time t*, without taking into account any later customer arrivals. This has been illustrated in Figure 3.6. Note that for every new arrival, $U(t)$ would have a step corresponding to the amount of work required by that arrival. At other times, whenever $U(t)>0$, it decreases at a unit rate. If $U(t)$ reaches zero, it stays at that value until the next arrival forces it to jump up by the required step size.

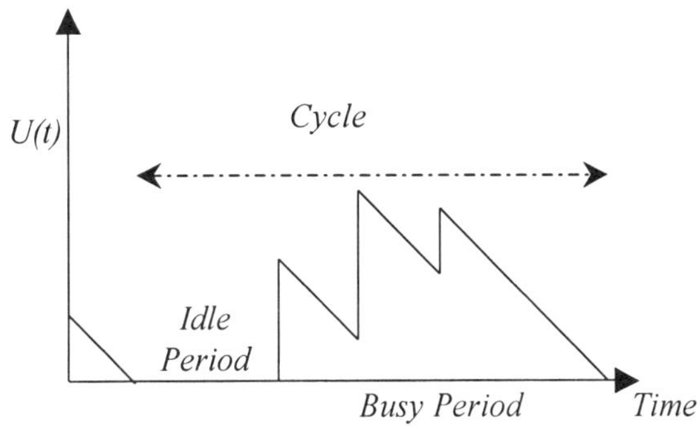

Figure 3.6. Unfinished Work in a M/G/1 Queue

The busy period may then be viewed as the time interval starting with the first arrival to an empty queue (i.e. ending the idle period) which continues as long as the unfinished work $U(t)$ stays non-zero (i.e. the queue does not go empty). The busy period ends and the next idle period begins, when the unfinished work becomes zero, i.e. the queue becomes empty. We let BP be the random variable denoting the length of such a busy period with probability density function $f_{BP}(t)$ whose L.T. is $L_{BP}(s)$. Similarly, the *Idle Period* starts with the departure of the customer who leaves behind an empty queue and ends with the arrival of the first subsequent customer - who would then start the next busy period. Let $f_{IP}(t)$ denote the probability density function of an idle period with length IP with L.T. $L_{IP}(s)$. Note that the random variables BP and IP will be independent of each other. The length of a *cycle* will then be the sum of the random variables BP and IP. Note that the cycle will also be of random length.

3.4.1 Idle Period (IP)

Since the arrival process is Poisson with rate λ, the inter-arrival times have an exponential distribution with mean $1/\lambda$. Using this, we get

$$\begin{aligned} P\{IP \geq y\} &= P\{\text{no arrival in an interval of length } y\} \\ &= e^{-\lambda y} \end{aligned}$$

for $y \geq 0$. Therefore the cumulative distribution function of the idle period will be $(1- e^{-\lambda y})$ and its probability density function will be $\lambda e^{-\lambda y}$ (i.e. an exponential distribution, as expected). The L.T. of its probability density function will then be

$$L_{IP}(s) = \frac{\lambda}{s+\lambda} \tag{3.17}$$

3.4.2 Busy Period (BP)

We obtain this based on the observation [Kle75] that the unfinished work $U(t)$ in the system *is invariant with respect to the service discipline* being followed in the queue. This effectively implies that once the customers are in the queue, we can permute the order in which they are served without changing the distribution of the busy period. This can be easily seen, as the busy period will end when the queue becomes empty once again regardless of the sequence in which the customers were being served during the busy period itself!

The implication of this observation is that we can assume the service discipline to be LCFS without changing the distribution of the busy period (even though the queue is/may actually be FCFS in nature.). We therefore consider the busy period for a LCFS queue in the following and will subsequently use the result obtained even for our FCFS queue. This is used to derive the probability density function $f_{BP}(t)$ and its L.T. $L_{BP}(s)$.

Consider a busy period that starts with the arrival of customer A_1. We assume the following –

(a) X_1 is the service time for A_1

(b) n^* arrivals (A_2,................., A_{n^*+1}) arrive during the service time X_1, in the sequence A_2,................., A_{n^*+1}.

(c) BP is the length of the busy period started by A_1 (whose distribution we want to find)

Consider the n^* arrivals (A_2,................., A_{n^*+1}) during the service time $X_1 = x_1$ of A_1. Since the service is LCFS in nature, each one of these will initiate *sub-*

busy periods BP_2,................, BP_{n^*+1} of its own so that we have the overall busy period as the sum of these sub-busy periods and the service time of A_1.

$$BP = X_1 + BP_2 + ...+ BP_{n^*+1} \qquad BP_j \perp BP_k \text{ and } BP_j \perp X_1 \quad \forall j,k$$

where X_1, BP_2,......., BP_{n^*+1} and n^* are all random variables. The following important observations can be made regarding the sub-busy periods and the busy period -

(a) The sub-busy periods are *independent and identically distributed* (iid) random variables.

(b) The distribution (cumulative distribution function, probability density function and L.T. of probability density function) of a sub busy period will be the same as the distribution of the (main) busy period.

We then get that

$$E\{e^{-s(BP)} \mid X_1 = x, n^* = k\} = e^{-sx}[L_{BP}(s)]^k$$

We first remove the conditioning on n^*, using the fact that these are arrivals over an interval x from a Poisson process with rate λ. This yields

$$\begin{aligned} E\{e^{-s(BP)} \mid X_1 = x\} &= e^{-sx}\sum_{k=0}^{\infty} e^{-\lambda x}\frac{(\lambda x)^k}{k!}[L_{BP}(s)]^k \\ &= e^{-sx}e^{-\lambda x}e^{\lambda x L_{BP}(s)} \\ &= e^{-x(s+\lambda-\lambda L_{BP}(s))} \end{aligned} \tag{3.18}$$

We can now remove the conditioning on X_1 using its probability density function $b(t)$ with L.T. $L_B(s)$ to get

$$L_{BP}(s) = E\{e^{-s(BP)}\} = \int_{x=0}^{\infty} e^{-x[s+\lambda-\lambda L_{BP}(s)]}b(x)dx$$

or $$L_{BP}(s) = L_B\left(s + \lambda - \lambda L_{BP}(s)\right) \tag{3.19}$$

Note that if we are given the average arrival rate λ of the Poisson Arrival process, and the probability density function $f_B(t)$ (or its L.T. $L_B(s)$) of the service times, then Eq. (3.19) may be solved to find $L_{BP}(s)$ of the busy period. From this, its probability density function $f_{BP}(t)$ may then be determined by taking the inverse Laplace transform of $L_{BP}(s)$. For very simple systems, it may actually be possible to solve this functional equation directly. It would be more common to use some kind of numerical recursion to solve it. However, even this may be computationally complex. The functional expression may of course be directly used to find the moments $E\{(BP)^k\}$ of the busy period. Some typical values obtained in this fashion are given below for $\rho=\lambda/\mu$.

$$E\{(BP)\} = \frac{\overline{X}}{(1-\rho)} \quad E\{(BP)^2\} = \frac{\overline{X^2}}{(1-\rho)^3} \quad E\{(BP)^3\} = \frac{\overline{X^3}}{(1-\rho)^4} + \frac{3\lambda(\overline{X^2})^2}{(1-\rho)^5}$$

$$E\{(BP)^4\} = \frac{\overline{X^4}}{(1-\rho)^5} + \frac{10\lambda\overline{X^2}\,\overline{X^3}}{(1-\rho)^6} + \frac{15\lambda^2(\overline{X^2})^3}{(1-\rho)^7} \quad \text{where } \overline{X^k} = E\{X^k\}$$

Note once again that the above results for the busy period of a M/G/1 queue will hold regardless of the service discipline followed by the queue. This follows once again from the argument that the distribution of the busy period itself will be invariant to the service discipline (as long as work is conserved and the server does not idle with a non-empty queue) and therefore, so will be its moments.

3.5 Delay Analysis for a LCFS M/G/1 Queue

In Section 3.3, we had done the delay analysis for the FCFS M/G/1 queue to obtain the distributions of the total time spent in the system by an arriving customer and the queueing delay encountered by a job (before its service is started). In this section, we consider the delay analysis for a LCFS M/G/1 queue to obtain the same distributions. Note that the distributions of the delay will depend on the service discipline even though the means will remain the same.

For this derivation, we consider a customer A arriving to the system. Two cases are possible here. The first case is when customer A arrives to an empty queue, with probability $(1-\rho)$; in this case, the queueing delay is zero and service to customer A starts immediately. The second case is more complicated and arises when customer A arrives to a non-empty queue (with probability ρ).

This second case has been shown in Figure 3.7. We consider this in more detail. Note that in this case, the total time T spent in the system by Customer A will be the sum of its queueing time Q and its service time X, i.e. $T = Q + X$. (In the first case, the queueing time would be zero as service would start immediately for customer A, immediately upon arrival.)

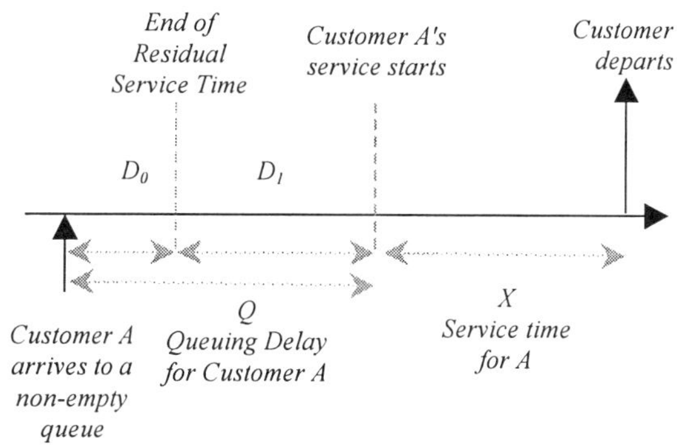

Figure 3.7. Customer Arrival and Departure from a LCFS M/G/1 Queue

Note that when it arrives to a non-empty queue, customer A will actually come in the middle of the ongoing service time of whichever customer is currently being served. Let D_0 be the residual service time for this service to complete. The Queueing Delay Q is shown to consist of D_0 and another component D_1. Since the queue service is LCFS, this second component D_1 will consist of *sub busy periods*, one associated with each of the customer arrivals in D_0. (Since the queue is LCFS, these will have to be served first before Customer A can be served.) Note that *D_0 and D_1 are not independent of each other*. This is evident as the length of D_0 affects the number arriving in it, which in turn affects the length D_1.

When customer A arrives to a non-empty queue, it will essentially sample an ongoing service where D_0 is the corresponding residual service time. From our earlier results on residual life, as given in Eq. (3.7), we can then state that

$$f_{D_0}(t) = \frac{1 - B(t)}{\overline{X}} \tag{3.20}$$

where *B(t)* is the cumulative distribution function of the service time X whose mean is $\overline{X}$. Taking transforms, we get

$$L_{D_0}(s) = \frac{1 - L_B(s)}{s\overline{X}} \tag{3.21}$$

Let N_0 be the number of arrivals in D_0. Then, for customer A coming to a non-empty queue, we get

$$E\left\{e^{-sD_1} \mid D_0 = y, N_0 = n\right\} = [L_{BP}(s)]^n$$

$$\& \quad E\left\{e^{-sQ} \mid D_0 = y, N_0 = n\right\} = e^{-sy}[L_{BP}(s)]^n$$

where $L_{BP}(s)$ is the L.T. of the probability density function of the busy period which can be found using Eq. (3.19). Therefore for the case where customer A arrives to a non-empty queue, we will have

$$E\left\{e^{-sD_1} \mid D_0 = y\right\} = \sum_{n=0}^{\infty} \frac{(\lambda y)^n}{n!} e^{-\lambda y}[L_{BP}(s)]^n = \exp[-y\{\lambda - \lambda L_{BP}(s)\}]$$

and

$$E\left\{e^{-sQ} \mid D_0 = y\right\} = \sum_{n=0}^{\infty} \frac{(\lambda y)^n}{n!} e^{-\lambda y} e^{-sy}[L_{BP}(s)]^n$$
$$= \exp[-y\{s + \lambda - \lambda L_{BP}(s)\}]$$

Using the probability density function (and its L.T.) of D_0, from Eqs. (3.20) and (3.21), we get that for the case of customer A coming to a non-empty queue, we will have

$$E\left\{e^{-sD_1}\right\} = \frac{1 - L_B(\lambda - \lambda L_{BP}(s))}{\overline{X}(\lambda - \lambda L_{BP}(s))} \tag{3.22}$$

and

$$E\left\{e^{-sQ}\right\} = \frac{1 - L_B(s + \lambda - \lambda L_{BP}(s))}{\overline{X}(s + \lambda - \lambda L_{BP}(s))} = \frac{1 - L_{BP}(s)}{\overline{X}(s + \lambda - \lambda L_{BP}(s))} \tag{3.23}$$

Therefore, considering both the cases where Customer A finds the queue empty and non-empty, we get

$$L_Q(s) = (1 - \rho) + \rho \frac{1 - L_{BP}(s)}{(s + \lambda - \lambda L_{BP}(s))\overline{X}} \tag{3.24}$$

as the L.T. of the probability density function of the queueing delay Q for the LCFS M/G/1 queue.

The L.T. of the probability density function of the overall delay T for the LCFS M/G/1 queue may also be found. This follows from the fact that $T=Q+X$ where X is the service time and that Q and X will be independent of each there. Therefore, we have

$$L_T(s) = L_Q(s) L_B(s) \tag{3.25}$$

Note that $L_{BP}(s)$ may be obtained by first solving Eq. (3.19). This may be used in Eq. (3.24) to get $L_Q(s)$. Since $L_B(s)$ is given (as it is the L.T. of the probability density function of the service time X), we can then use Eq. (3.24) and Eq. (3.25) to obtain $L_T(s)$.

3.6 The M/D/1 Queue

The M/D/1 queue has Poisson arrivals like the M/G/1 queue, but the service times are fixed (i.e. deterministic). Study of this kind of a queue has become more important because this may be a convenient way to model an *Asynchronous Transfer Mode* (ATM) node with fixed size cells as the jobs requiring service. This may also be a good model for a packet switching node or router in a computer network where the packets are of fixed size. The results obtained for the M/G/1 queue in Sections 3.1-3.5 may be easily applied for this queue to obtain the corresponding results.

Let m be the duration of service (which is fixed). We will then have

$$\overline{X} = m, \ \overline{X^2} = m^2, \ \rho = \lambda m, \ L_B(s) = e^{-sm}$$

We can use Eqs. (3.1) and (3.2), respectively, to get the residual service time R and the mean queueing delay W_q.

$$\begin{aligned} R &= \frac{1}{2}\lambda m^2 \\ W_q &= \frac{\lambda m^2}{2(1-\rho)} = \frac{m}{2}\frac{\rho}{1-\rho} \end{aligned} \tag{3.26}$$

Knowing the mean queueing delay W_q, the mean time W spent in the system may be found as

$$W = m + W_q = \frac{m(2-\rho)}{2(1-\rho)} \tag{3.27}$$

Similarly, Eq. (3.14) may be used to get the generating function *P(z)* for the number in the system.

$$P(z) = \frac{(1-\rho)(1-z)e^{-m\lambda(1-z)}}{e^{-m\lambda(1-z)} - z} = \frac{(1-\rho)(1-z)e^{-\rho(1-z)}}{e^{-\rho(1-z)} - z} \qquad \rho = m\lambda \tag{3.28}$$

It should be noted that the mean results W and W_q and the generating function *P(z)* will hold regardless of the service discipline followed by the M/D/1 queue, i.e. whether it is FCFS, LCFS, SIRO etc..

For the FCFS M/D/*1* queue, we can use $P(z)= L_T(\lambda-\lambda z)$ as in Section 3.3, to find the L.T. $L_T(s)$ of the probability density function $f_T(t)$ of the time T spent in system by an arriving customer. This is then given by

$$L_T(s) = \frac{s(1-\rho)e^{-sm}}{s - \lambda + \lambda e^{-sm}} \qquad \text{(FCFS M/D/1 Queue)} \tag{3.29}$$

This Laplace Transform may be inverted to find the actual probability density function $f_T(t)$ (at least, we can do that in principle, even if the actual inversion is not easy to do). We can also use the transform of Eq. (3.29) directly to find the various moments of the time spent in system by a customer arriving to this queue. One can also use this to find the distribution (probability density function or its L.T.) or the moments of the time spent waiting in the FCFS M/D/1 queue by an arriving customer. This may be obtained from $L_Q(s)=L_T(s)/L_B(s)$ using the fact that the queueing delay Q and the service time X are independent random variables.

The busy period distribution of the M/D/1 queue may also be found by using Eq. (3.19), with proper substitutions in the M/G/1 results. This gives

$$L_{BP}(s) = e^{-m(s+\lambda-\lambda L_{BP}(s))} = e^{-ms-\rho}e^{\rho L_{BP}(s)} \tag{3.30}$$

Note that $L_{BP}(s)$ may then be obtained by solving the equation -

$$L_{BP}(s)[\exp(-\rho L_{BP}(s))] = e^{-ms-\rho} \tag{3.31}$$

For the LCFS *M/G/1* queue, one can similarly obtain the distribution of the queueing delay from Eq. (3.25) by proper substitution. This will be

$$L_Q(s) = (1-\rho) + \frac{\lambda[1 - L_{BP}(s)]}{s + \lambda - \lambda L_{BP}(s)} \tag{3.32}$$

using $L_{BP}(s)$ as obtained earlier in Eq. (3.30).

3.7 Alternative Derivation for the Delays in a FCFS M/G/1 Queue

We can re-derive our earlier results of Section 3.3 on the distribution of the queueing delay and total delay for a FCFS M/G/1 queue following an alternative approach given in this section. Though the results obtained are the same, the approach given here illustrates a direct, but somewhat more difficult way of obtaining the same results. The derivation of Section 3.3 was somewhat indirect as it was based on the observation that for a FCFS M/G/1 queue, the number left behind in the system by a departing customer will be the number arriving while this customer was in the system. This observation was used to derive the distributions of the total delay and the queueing delay.

The derivation given in this section actually looks at how the busy period evolves in a FCFS M/G/1 queue. For this, we decompose a busy period of the M/G/1 (FCFS) queue into a sequence of *dependent* random variables X_0, X_1, X_2......... as follows.

X_0 service time (X) of the first customer (say Customer 1) initiating the busy period
X_1 time taken to serve all customers arriving in X_0
X_2 time taken to serve all customers arriving in X_1
... .
... .
X_i time taken to serve all customers arriving in X_{i-1}
... .
... .
etc.

The sequence X_0, X_1, X_2 .... terminates with X_N, if no customers arrive to the queue during X_N. In that case, X_N will be the terminating interval in the sequence and $X_{N+1}=X_{N+2}=.........X_\infty=0$. It is possible to show that if $\rho<1$, then the sequence $\{X_i\}$ will indeed terminate and therefore the busy period will be of finite length.

The *Busy Period Duration B* will then be given as

$$B = \sum_{i=0}^{\infty} X_i \tag{3.33}$$

with the understanding that the intervals X_i beyond the terminating one are of zero length. Let $L_{X(i)}=E\{exp(-sX_i)\}$ be the L.T. of the probability density function of the random variable X_i and let n_i represent the number of arrivals (random) in X_i. We can then derive the following by successively removing the conditioning terms one by one.

$$E\{e^{-sX_i} \mid X_{i-1} = x, n_{i-1} = n\} = [L_B(s)]^n$$

$$E\{e^{-sX_i} \mid X_{i-1} = x\} = \sum_{n=0}^{\infty} \frac{(\lambda x)^n}{n!} e^{-\lambda x} L_B^n(s) = e^{-\lambda x(1-L_B(s))}$$

$$E\{e^{-sX_i}\} = \int_{x=0}^{\infty} e^{-\lambda x(1-L_B(s))} dX_{i-1}(x)$$

Therefore

$$L_{X(i)}(s) = L_{X(i-1)}(\lambda - \lambda L_B(s)) \tag{3.34}$$

This gives the way the distributions of the successive time intervals in $\{X_i\}$ will evolve within a busy period.

Consider now a particular arrival, say customer A, which comes during the busy period. Assume that this arrival comes during the interval X_i and that it encounters a queueing delay of Q^*. Therefore, we have

Q^* = Residual Life of the interval X_i + $\sum$(service times of all those customers who arrived in the interval X_i but before customer A)

Let Y_i be the residual life of the interval X_i when customer A arrives and let N_i be the number of arrivals in X_i which came before customer A. Then -

$$E\{e^{-sQ^*} \mid i, X_i = \alpha, Y_i = \beta, N_i = n\} = e^{-s\beta} L_B^n(s) \tag{3.35}$$

Removing the conditioning on N_i using the Poisson distribution of the arrivals, we get

$$E\{e^{-sQ^*} \mid i, X_i = \alpha, Y_i = \beta\} = e^{-s\beta} \sum_{n=0}^{\infty} \frac{[\lambda(\alpha-\beta)]^n}{n!} e^{-\lambda(\alpha-\beta)} L_B^n(s)$$
$$= e^{-s\beta - \lambda(\alpha-\beta) + \lambda(\alpha-\beta) L_B(s)} \tag{3.36}$$

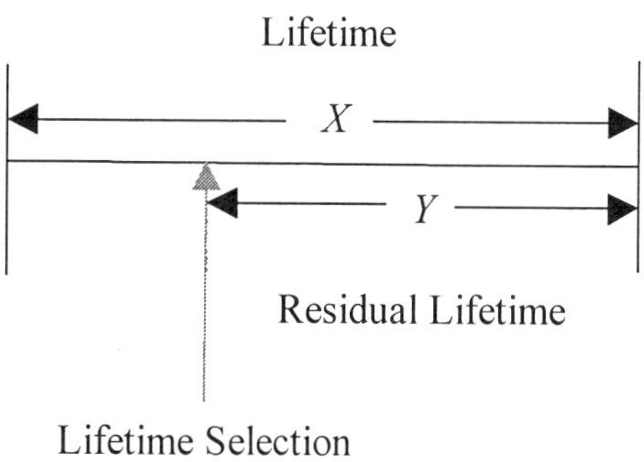

Figure 3.8. Actual Lifetime and Residual Lifetime

Recall that if X is a life time and Y is a residual life time as shown above in Figure 3.8, then Y will always be less than X. For $y \le x$, we have from Eqs. (3.5) and (3.6) that

$$P\{Y \le y \mid X \le x\} = \frac{y}{x}$$

$$P\{y \le Y \le y + dy, x \le X \le x + dx\} = \frac{f_X(x)}{E\{X\}} dxdy$$

Using this in Eq. (3.36), we get that

$$E\{e^{-sQ^*} \mid i\} = \int_{\alpha=0}^{\infty} \int_{\beta=0}^{\alpha} e^{-[s-\lambda+\lambda L_B(s)]\beta} e^{-[\lambda-\lambda L_B(s)]\alpha} \frac{f_{X_i}(\alpha)d\alpha}{E\{X_i\}} d\beta$$
$$= \int_{\alpha=0}^{\infty} \frac{e^{-s\alpha} - e^{-[\lambda-\lambda L_B(s)]\alpha}}{-s+\lambda-\lambda L_B(s)} \frac{f_{X_i}(\alpha)d\alpha}{E\{X_i\}}$$
$$= \frac{E\{e^{-sX_i}\} - E\{e^{-\lambda[1-L_B(s)]X_i}\}}{E\{X_i\}[-s+\lambda-\lambda L_B(s)]}$$

and therefore,

$$E\{e^{-sQ^*} \mid i\} = \frac{L_{X(i)}(s) - L_{X(i)}(\lambda - \lambda L_B(s))}{E\{X_i\}[-s + \lambda - \lambda L_B(s)]} \tag{3.37}$$

Using Eq. (3.34) in (3.37), we obtain that

$$E\{e^{-sQ^*} \mid i\} = \frac{L_{X(i+1)}(s) - L_{X(i)}(s)}{E\{X_i\}(s - \lambda + \lambda L_B(s))} \tag{3.38}$$

Given that the arrival of Customer A happened during a busy period (of length *BP*), the probability that it will occur during the ith such interval X_i will be $E\{X_i\}/E\{BP\}$. This may then be used to derive

$E\{e^{-sQ^*} \mid$ customer A arrives in busy period$\}$

$$\begin{aligned} &= \sum_{i=0}^{\infty} E\{e^{-sQ^*} \mid i\} \frac{E\{X_i\}}{E\{BP\}} \\ &= \frac{1}{E[BP\}[s - \lambda + \lambda L_B(s)]} \sum_{i=0}^{\infty} [L_{X(i+1)}(s) - L_{X(i)}(s)] \\ &= \frac{1 - L_B(s)}{E[BP\}[s - \lambda + \lambda L_B(s)]} \end{aligned} \tag{3.39}$$

Note that the last expresion follows from the fact that

$$L_{X(0)}(s) = E\{e^{-sX_0}\} = E\{e^{-sX}\} = L_B(s)$$

since X_0 is just a service time X and $L_{X(\infty)}(s) = 1$ since $X_\infty = 0$ beyond the busy interval.

We can also obtain the mean busy period length, $E\{BP\}$, from simple arguments. Consider an idle period *IP* (of mean leangth $1/\lambda$), followed by a busy period such that the sum of the two will constitute a cycle. Then, we have

$$\rho = \frac{E\{BP\}}{E\{BP\} + E\{IP\}} = \frac{E\{BP\}}{E\{BP\} + \lambda^{-1}} \tag{3.40}$$

This may be simplified to give

$$E\{BP\} = \frac{E\{X\}}{(1-\rho)}$$

which leads to

$$E\{e^{-sQ^*} \mid \text{customer A arrives in busy period}\} = \frac{(1-\rho)(1-L_B(s)}{E\{X\}(s-\lambda+\lambda L_B(s))}$$

Considering together the two cases where customer A arrives in an idle period and when it arrives in a busy period, we get

$$L_Q(s) = (1-\rho) + \rho E\{e^{-sQ^*} \mid \text{customer A arrives in busy period}\}$$

This may be simplified to get our earlier result that

$$L_Q(s) = \frac{s(1-\rho)}{s-\lambda+\lambda L_B(s)}$$

as given in Eq. (3.16) for the FCFS M/G/1 queue.

Problems

1. For a particular M/G/1 queue, the Laplace Transform of the service time is given to be

$$L_B(s) = \frac{0.5s(\mu_1 + \mu_2) + \mu_1\mu_2}{(s + \mu_1)(s + \mu_2)}$$

Analyse this queue to the extent possible, using both the residual life approach and the imbedded Markov chain approach.

2. Consider a special M/G/1 queue, which operates normally except when the system becomes empty. Whenever that happens, service starts again only when there are K customers present in the system, where K is a given constant. Once service begins, it continues normally until the system becomes empty once again. For this queue, obtain the following.

(a) The probability that the system is empty

(b) The distribution for the number in the system

(c) The mean number in the system

(d) The mean queueing delay

3. Consider a special M/G/1 queue operating in the following fashion. A new customer (say customer A) entering the queue starts service *immediately* regardless of whether the queue is empty or non-empty when it arrives. In the latter case, it pre-empts the customer currently being served (say customer B) who is pushed back into the queue once again. Service to the pre-empted customer B *resumes from its point of interruption* after service to customer A and service to all subsequent arrivals after customer A get completed. [Note that work is still conserved as service is interrupted on a pre-emptive resume basis.]

(a) For this queue, show that the distribution (L.T. of the probability density function) of the time an arriving customer will spend in the system will be given by $L_G(s)=L_B(s+\lambda-\lambda L_G(s))$ with mean $\frac{\overline{X}}{(1-\lambda\overline{X})}$.

Note that this time effectively has the same distribution as the busy period of the normal M/G/1 queue.

(b) The distribution (generating function) of the number of other customers served while a particular customer is in the queue will be given by $F(z) = L_G(\lambda - \lambda z)$ with mean $\dfrac{\lambda \overline{X}}{(1-\lambda \overline{X})}$.

4. Consider a special M/G/1 queue where a customer arriving tosses a coin that has probability p of coming up heads and probability $(1-p)$ of coming up tails. If the coin shows heads, then the customer leaves without service (i.e. service time is zero). If it comes up tails, then the customer goes for a service time that is exponentially distributed with mean $1/\mu$. Analyse this queue, to the extent possible using the approaches given in this section both for the case where the service discipline is FCFS and for the case where it is LCFS.

5. Consider a M/G/1 queue where the service times have the probability density function $f_X(t)$ and L.T. $L_B(s)$. Whenever a customer finishes service, it either decides to go for another service immediately with probability p (and this may continue subsequently, as well) or leaves the queue with probability $(1-p)$. Analyse this queue for the two cases where the service disciplines are FCFS and LCFS in nature.

One can also consider a variation of this queue, where a customer who on finishing service decides to go for another service joins the queue once again as a new customer would. Consider this for the LCFS and FCFS variations as well.

Chapter 4

Advanced Queueing Theory

Vacations, Bulk Arrivals and Priorities in a M/G/1 Queue and the Geo/G/1 Queue

In this chapter, we consider several useful variations of the M/G/1queue, the basic model of which was discussed in Chapter 3. These variations are useful as they may be used to model situations where, even though the basic queue is a single-server one with an infinite buffer, several other special considerations apply. The analytical approaches used for these queueing models are similar to those used in Chapter 3 but require some special techniques, which are interesting and useful to know.

We first consider *queues with vacations* where the server stops serving for some time even though there may be users still in the queue who are waiting for service. Note that this will not happen in a simple M/G/1 queue, as regardless of the queueing discipline followed, the server will never be allowed to idle unless the system itself is empty. Two different ways of modelling such vacations have been considered. We also consider the special case *exceptional first service* to a M/G/1 queue where the first customer starting a busy period requires some special service different from the way the other subsequent customers in that busy period are served. This is commonly observed in several real life situations. For example, in a computer network, the first packet of a multi-packet transmission may require special handling because of the buffer allocations and routing decisions that are required to be made. The other packets of this multi-packet transmission may then be served without these special considerations. This chapter also considers queues with batch arrivals where the batches arrive from a Poisson process but a batch may have one or more (random) customers. This is similar to the batch arrivals considered in Section 2.10 except that there we were limited to only exponentially distributed service times. We also describe priority queues and analyse single-server M/G/1 queues for different priority mechanisms. The chapter concludes with a description and analysis of discrete-time queues of the type Geo/G/1 and $Geo^{[X]}/G/1$.

4.1 M/G/1 Queue with Vacations

Consider a M/G/1 queue, where after each busy period, the server goes on a *vacation* for a random interval of time. The server is not available for providing service while it is on vacation. Arrivals coming during the vacation can go into service only after the server returns from vacation. This type of behaviour is exhibited in many real life situations. This, for example, will be the case if a server serving a queue goes for a coffee break whenever the system becomes empty and can resume service only after it returns from its coffee break. Scenarios like this may also arise in communication systems and other service models. For example, in a polling type situation, a single server may provide service at a number of service locations moving from one to other in some pre-defined sequence. One way of modelling such a system would be to assume that from the service station's point of view, the server goes for a vacation (i.e. moves to the other stations) after completing service to all the customers at its present location. Service at the service station being considered resumes only after the server returns to this station. Note that polling systems of this type may also be modelled in other way, not necessarily invoking the vacation model considered here [BeG92].

We assume in this section that if on returning from a vacation, the server still finds the system empty, then it goes on another vacation of random length. This continues until the server returns from the vacation and finds jobs/customers waiting in the queue – it then starts serving customers normally once again. After it starts service, following a vacation, the server continues serving normally (like a normal M/G/1 queue) until the system becomes empty once again. At that time, the server once again leaves for a random length vacation and the process is repeated.

This would be the basic vacation model that we will discuss here. Simple variations to this are also encountered. Some examples are -

(a) Server goes on *only one vacation* after the busy period ends. Once it comes back from this vacation, it does not go for another vacation even if the system is still empty at that time. This has been discussed in Section 4.2

(b) On return from a vacation, server starts service only if it finds *K or more* jobs/customers waiting in queue. If the number waiting is less than *K*, then it goes for another vacation. [Note that our basic vacation model described above and analysed subsequently is really the case for $K=1$.]

As for the case of the normal M/G/1 queue, we will analyse this queue using both the (a) *Residual Life-Time Approach* and the (b) *Imbedded*

Markov Chain Approach. Note that the analyses for the variations to the basic vacation model may also be done using either of these two approaches.

4.1.1 Residual Life Approach

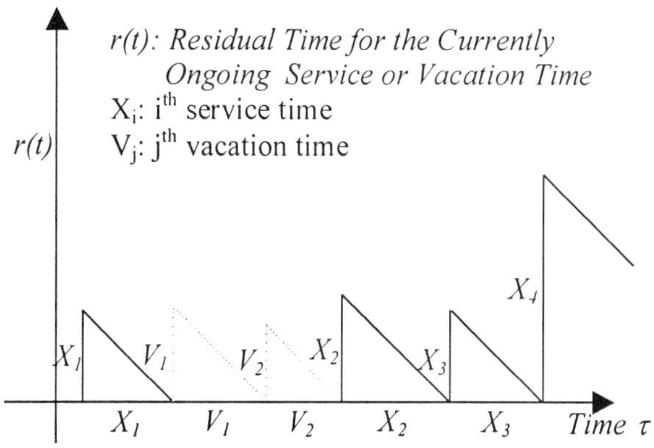

Figure 4.1. Residual Time $r(\tau)$ for a M/G/1 Queue with Vacations

For this we follow the same approach as for the simple M/G/1 queue analysed in Section 3.1. As for the simple M/G/1 queue, we draw the plot of the residual life-time $r(t)$ as a function of time t as shown in Figure 4.1. In this case, however, the residual life time measures either the time left to the end of the current service or the current vacation, depending on whether a service or a vacation is on-going at time t. Let X_i be the ith service time and V_j the jth vacation interval.

We consider the interval $(0, t)$ and evaluate the mean value of $r(t)$ over this interval as

$$\text{Time Average of } r(t) \text{ over } (0, t) = \frac{1}{t}\int_0^t r(x)dx = \frac{1}{t}\sum_{i=1}^{M(t)}\frac{1}{2}X_i^2 + \frac{1}{t}\sum_{j=1}^{L(t)}\frac{1}{2}V_j^2$$

where $M(t)$ is the number of arrivals in the interval $(0,t)$ and $L(t)$ is the number of vacation intervals in that same interval. Note that is really an approximation ignoring minor errors in the last residual time (either service time or vacation time, as the case may be), as was justified for the case in Section 3.1. This error will tend to zero as $t \to \infty$. Using the definitions of $M(t)$ and $L(t)$, we get that

$$\frac{1}{t}\int_0^t r(x)dx = \frac{1}{2}\frac{M(t)}{t}\frac{1}{M(t)}\sum_{i=1}^{M(t)} X_i^2 + \frac{1}{2}\frac{L(t)}{t}\frac{1}{L(t)}\sum_{j=1}^{L(t)} V_j^2 \tag{4.1}$$

Taking appropriate limits of the terms above as $t \to \infty$, we get

$$R = \lim_{t\to\infty} \frac{1}{t}\int_0^t r(x)dx \qquad \lim_{t\to\infty} \frac{M(t)}{t} = \lambda$$

$$\lim_{t\to\infty} \frac{1}{M(t)}\sum_{i=1}^{M(t)} X_i^2 = \overline{X^2} \qquad \lim_{t\to\infty} \frac{1}{L(t)}\sum_{j=1}^{L(t)} V_j^2 = \overline{V^2}$$

Note that as before, $\overline{X}$ and $\overline{X^2}$ are the first and second moments, respectively of the service times. For the vacation intervals, $\overline{V}$ and $\overline{V^2}$ are the first and second moments, respectively. We note also that for $t \to \infty$, the fraction of time the server will stay idle will tend to *(1-ρ)* where ρ is the offered traffic $\rho = \lambda E\{X\} = \lambda \overline{X}$. Since *t(1-ρ)* will be the total time interval in *(0,t)* over which the server will be idle, therefore $\lim_{t\to\infty} \frac{t(1-\rho)}{L(t)} = \overline{V}$. Using these in conjunction with Eq. (4.1), we get that for this case the mean residual time R will be given by

$$R = \frac{1}{2}\lambda\overline{X^2} + \frac{1}{2}(1-\rho)\frac{\overline{V^2}}{\overline{V}} \tag{4.2}$$

However, as in the case of the simple *M/G/1* queue, the mean queueing delay W_q may be expressed as the sum of the mean residual time R and the mean time spent to serve the N_q customers ahead of the new arrival who are already in the queue. (Note that though this implicitly assumes a FCFS queue, this will actually be true for any service discipline.) We can then say that

$$W_q = N_q\overline{X} + R = \lambda W_q \overline{X} + R$$

Defining $\rho = \lambda E\{X\} = \lambda\overline{X}$, and substituting the mean residual time R from Eq. (4.2), we get the mean queueing delay for this case as

$$W_q = \frac{\lambda \overline{X^2}}{2(1-\rho)} + \frac{\overline{V^2}}{2\overline{V}} \tag{4.3}$$

The above analysis was based on the assumption that the intervals $\{X_i\}$ and $\{V_i\}$ are independent of each other and are also independent of the arrival process. Actually, these conditions may be relaxed to some extent and strict independence of this type is not really needed for Eq. (4.3) to hold [BeG92].

4.1.1 Imbedded Markov Chain Approach

For this, we imbed our Markov Chain of system states (denoting the number in the system) at the time instants t_i $i=1, 2, 3,$ .... when the ith customer departs from the system. As in Section 3.2 and [Kle75], at the time instant t_i, the system state n_i is defined to be the number of customers left behind in the system when the ith customer departs. Similarly, let a_{i+1} be the number of arrivals in the $(i+1)^{th}$ service time. For the M/G/1 queue with vacations, we additionally define that

j = number of customers waiting for service when a busy period begins, $j \geq 1$

and f_j = P{j customers starting the busy period}

Note that these additional terms j and f_j are needed because one or more customers may come during the last vacation interval after which the busy period begins. Therefore, the busy period will start with either one customer or with more than one customer.

The generating function $F(z)$ of the number of customers waiting in queue when the busy period starts is given by

$$F(z) = \sum_{j=1}^{\infty} f_j z^j = E\{z^j\}$$

We also define $f_V(t)$ to be the probability density function of the vacation interval with cumulative distribution function $F_V(t)$ and L.T. (of the probability density function) as $L_V(s)$. For the imbedded time points t_i at the customer departure instants, we then can write that

$$n_{i+1} = n_i + a_{i+1} - 1 + j[1 - U(n_i)] \qquad \text{for } i=1, 2, ,3, ... \tag{4.4}$$

where $U(n_i)$ is the unit step function. Note that Eq. (4.4) may also be explicitly written as

$$\begin{aligned} n_{i+1} &= a_{i+1} + j - 1 && \textit{for } n_i = 0 \\ &= n_i + a_{i+1} - 1 && \textit{for } n_i \geq 1 \end{aligned} \tag{4.5}$$

At steady state, the dependence on i may be dropped. Further, we define $A(z)$ to be the generating function of the number arriving in a service time. Since PASTA will be applicable to this queue (the arrival process is Poisson) and Kleinrock's Result is also applicable (see Section 3.2), the probability p_n of there being n users in the system will be the same for both the arrival and the departure instants. Moreover, these same probabilities will also hold for the ergodic situation when the system is in steady state. Using these and simplifying, we get

$$\begin{aligned} P(z) &= E\{z^{n+a-1+j[1-U(n)]}\} = E\{z^a\}E\{z^{n-1+j[1-U(n)]}\} \\ &= A(z)E\{p_0 z^{j-1} + \sum_{n=1}^{\infty} z^{n-1} p_n\} \\ &= A(z)\left[p_0 E\{z^{j-1}\} + \sum_{n=1}^{\infty} z^{n-1} p_n \right] \\ &= A(z)\left[\frac{p_0}{z} F(z) + \frac{1}{z} P(z) - \frac{1}{z} p_0 \right] \end{aligned}$$

Simplifying this yields the required generating function $P(z)$ as

$$P(z) = p_0 A(z) \frac{1 - F(z)}{A(z) - z} \tag{4.6}$$

To use Eq. (4.6), we need to know the terms p_0,, $A(z)$ and $F(z)$. Note that we would then need to apply the condition that $P(1)=1$ to find p_0 before we can give the actual expression for $P(z)$. We can also find p_0 by taking the mean of both the LHS and RHS of either Eqs. (4.4) or (4.5) under steady state conditions. This will be equivalent to obtaining this probability by applying the normalisation condition inherent in $P(1)=1$. We further note the following

(i) Since $A(z)$ is the generating function of the number arriving in a service time (probability density function $b(t)$ with L.T. $L_B(s)$) from a Poisson process at rate λ. From earlier derivations of Section 3.2, as in Eq. (3.13), we get that $A(z)=L_B(\lambda-\lambda z)$

(ii) Let $F^*(z)$ be the generating function of the number of arrivals in one vacation interval. Then, we would get as before that $F^*(z)=L_V(\lambda-\lambda z)$.

(iii) We can also show that

$$F(z)=\frac{F^*(z)-F^*(0)}{1-F^*(0)}=\frac{L_V(\lambda-\lambda z)-L_V(\lambda)}{1-L_V(\lambda)} \tag{4.7}$$

$$\frac{F''(1)}{F'(1)}=-\frac{\lambda L_V''(0)}{L_V'(0)}=\lambda\frac{\overline{V^2}}{\overline{V}} \tag{4.8}$$

$$A(1)=P(1)=F(1)=1$$

$$A'(1)=\rho=\lambda\overline{X} \qquad A''(1)=\lambda^2\overline{X^2} \qquad F'(1)=\frac{\lambda\overline{V}}{1-L_V(\lambda)}$$

where $\overline{V}$ and $\overline{V^2}$ are respectively the first and second moments of the vacation interval. We can now differentiate both sides of Eq. (4.6) and evaluate the LHS and RHS at $z=1$. Using the results given in Eq. (4.8) and evaluating provides us with the probability p_0 of the system being empty.

$$p_0=\frac{1-\rho}{F'(1)} \tag{4.9}$$

Substituting these in Eq. (4.6), we get the required generating function $P(z)$.

$$P(z) = \left[\frac{1-\rho}{F'(1)}\right] L_B(\lambda - \lambda z) \frac{\left(1 - \frac{L_V(\lambda - \lambda z) - L_V(\lambda)}{1 - L_V(\lambda)}\right)}{L_B(\lambda - \lambda z) - z}$$
$$= \frac{(1-\rho)}{\lambda \overline{V}}(1 - L_V(\lambda - \lambda z))\left(\frac{L_B(\lambda - \lambda z)}{L_B(\lambda - \lambda z) - z}\right) \quad (4.10)$$

Note that this holds for the customer departure instants from the M/G/1 queue. However, since Kleinrock's result will hold for this queue, this generating function will also hold for the customer arrival instants. Since the arrival process is Poisson, PASTA would also apply. Hence, the above generating function would also hold for the time averaged system state at equilibrium.

The *P(z)* obtained in Eq. (4.10) may be used to find the mean number in the system *N*. We can then use Little's Result on this to get the mean time spent in system *W* as N/λ. From this, we can obtain the mean time spent waiting in queue as $W_q = W - E\{X\}$. This can be shown to be

$$W_q = \frac{\lambda \overline{X^2}}{2(1-\rho)} + \frac{\overline{V^2}}{2\overline{V}} \quad (4.11)$$

as obtained in Eq. (4.3) earlier by the Residual Life-Time Based Approach of Section 4.1.1.

One important observation should be noted about the results obtained in this section. For the results on the first moments (mean quantities W, W_q, N, and N_q), the queue may have any service discipline, i.e. FCFS, LCFS or SIRO. The results on the probability generating function (and hence the state probability distribution) of the number in the system also hold for any service discipline. The second and higher moments of the delay quantities and their distributions (i.e. probability density function or L.T. of probability density function) will depend on the nature of the actual service discipline that is being followed.

It should also be noted, that in this case, the server may be idle even when there are customers waiting. This, for example, may happen if at that time, the server is on a vacation. This was not allowed in the simple M/G/1 queue considered earlier. However, once the server starts serving (i.e. the busy period starts), the server can no longer idle until the queue becomes empty once again.

4.2 M/G/1 Queue with Only One Vacation after Idle

In this case, the server goes for only one vacation of random length when the queue becomes idle. (Note that for the case considered in Section 4.1, the server may go on multiple consecutive vacations provided it finds the system empty whenever it returns from a vacation.) Once this vacation is over, it remains in the system and waits for a customer to come (so that it can start service) even if it finds the queue empty when it returns from its vacation.

Let $f_V(t)$ to be the probability density function of the vacation interval with cumulative distribution function $F_V(t)$ and L.T. $L_V(s)$ (of the probability density function) as defined earlier in Section 4.1. Let j be the number in the system at the end of the vacation with distribution f_j and generating function $F(z)$. Note that the following will hold

$$F(z) = L_V(\lambda - \lambda z) \ , f_0 = F(0) = L_V(\lambda), \ F'(1) = \lambda \overline{V} \ , \ F''(1) = \lambda^2 \overline{V^2}$$

where $\overline{V}, \overline{V^2}$ are respectively the first and second moments of the vacation interval. If the *Imbedded Markov Chain* approach is used, we can shown that for this case the number left behind by the $(i+1)^{th}$ departure may be written as

$$\begin{aligned} n_{i+1} &= a_{i+1} + j - U(j) && \text{for } n_i = 0 \\ &= n_i + a_{i+1} - 1 && \text{for } n_i \geq 1 \end{aligned} \tag{4.12}$$

From this, following the same procedure as in Section 4.1.2, we can obtain the results that

$$p_0 = \frac{1 - \lambda \overline{X}}{\lambda \overline{V} + L_V(\lambda)} \tag{4.13}$$

and

$$P(z) = \left(\frac{1 - \lambda \overline{X}}{\lambda \overline{V} + L_V(\lambda)} \right) \left(\frac{L_B(\lambda - \lambda z)}{z - L_B(\lambda - \lambda z)} \right) \left(L_V(\lambda - \lambda z) - (1 - z) L_V(\lambda) - 1 \right) \tag{4.14}$$

The *Residual Life* approach of Section 4.1.1 may also be applied to analyse this queueing situation. Note that, as before, we will have

$$W_q = \frac{R}{1-\lambda\overline{X}}$$

However, the expression for the mean residual time R will differ from that of the previous case in Section 4.1.1. Note that in an interval $(0,t)$, the total mean idle time will be $t(1-\lambda E\{X\})$and the mean length of each idle period in a cycle will be given by

$$E\{IP\} = \overline{IP} = \left(\overline{V} + f_0 \frac{1}{\lambda}\right) = \left(\overline{V} + \frac{1}{\lambda} L_V(\lambda)\right)$$

Following the graphical approach given in Sec. 4.1.1, we can then derive that

$$R = \frac{\lambda \overline{X^2}}{2} + \frac{1-\lambda\overline{X}}{\left(\overline{V} + \frac{1}{\lambda} L_V(\lambda)\right)} \frac{\overline{V^2}}{2} \tag{4.14}$$

and

$$W_q = \frac{\lambda \overline{X^2}}{2(1-\lambda\overline{X})} + \frac{\overline{V^2}}{2\left(\overline{V} + \frac{1}{\lambda} L_V(\lambda)\right)} \tag{4.15}$$

Following our usual approach, once the mean queueing delay W_q is known, the mean number N_q waiting for service in the queue may be found by applying Little's result. One can also find the mean time W spent in the system knowing W_q and the mean service time. Little's result may be applied once again to W to find the mean number N in the system.

4.3 M/G/1 Queue with Exceptional First Service

We consider here a special kind of M/G/1 queue where the first customer to get serviced when a busy period starts gets a different kind of service (i.e. with a different distribution of the service time) than the other customers served during the busy period. This kind of situation may arise in many practical situations as the server starting work after a period of idling may work slower (or faster) than the way it otherwise would. In a computer network, this may also arise because the first packet in a sequence may require special processing for route establishment and buffer set-up and will therefore require a different kind of service than the subsequent packets.

Once again, it is possible to analyse this using either the residual lifetime approach or the method of imbedded Markov chain. In the following, the latter method will be described based on the choice of the customer departure instants $\{t_i\}$ as the imbedded time points and the system state represented by the number in the system $\{n_i\}$ as seen by a departing customer. Note that both PASTA and Kleinrock's Result will still be applicable to this queue allowing us to generalize the distributions obtained at the departure instants to both the arrival instants and the ergodic system averages.

Let $b(t)$ (with L.T. $L_B(s)$) be the probability density function of the normal service time and let $b^*(t)$ (with L.T. $L_{B^*}(s)$) be the probability density function of the service time of the first customer being served in a busy period, i.e. that of the exceptional first service time. In the following we will use superscript * to denote quantities relevant to the exceptional first service. For this system, we can then show that

$$n_{i+1} = n_i - U(n_i) + a^*_{i+1} + (a_{i+1} - a^*_{i+1})U(n_i) \tag{4.16}$$

where a^*_{i+1} is the number of arrivals in the first (exceptional) service time of the busy period and a_{i+1} is the number of arrivals in the normal service times. The generating functions for these will be denoted by $A^*(z)$ and $A(z)$, respectively and may be found using $A^*(z)=L_{B^*}(\lambda-\lambda z)$ and $A(z)=L_B(\lambda-\lambda z)$. Taking expectations of the RHS and LHS of Eq. (4.16) at equilibrium (i.e. by dropping the subscript i), we get that

$$1 - p_0 = \frac{\overline{a^*}}{1 - \bar{a} + \overline{a^*}} \quad \text{with } \bar{a} = \lambda \bar{X} \text{ and } \overline{a^*} = \lambda \overline{X^*} \tag{4.17}$$

Therefore,

$$p_0 = \frac{1 - \lambda \bar{X}}{1 - \lambda \bar{X} + \lambda \overline{X^*}} \tag{4.18}$$

Using Eq. (4.16), we can also find that

$$P(z) = \frac{p_0[A(z) - zA^*(z)]}{A(z) - z} \tag{4.19}$$

where $A^*(z)=L_{B^*}(\lambda-\lambda z)$ and $A(z)=L_B(\lambda-\lambda z)$.

It should be noted that the above results will hold even if the queue is not FCFS in nature. If a FCFS service discipline is assumed, we can find the distribution of the total time T spent in system by an arriving customer by identifying as before that $P(z)=L_T(\lambda-\lambda z)$. This would be the L.T. of the probability density function of T and will be given by

$$L_T(s) = \frac{p_0[\lambda L_B(s) - (\lambda - s)L_{B^*}(s)]}{\lambda L_B(s) + s - \lambda} \tag{4.20}$$

This needs to be inverted to find the actual distribution (probability density function) of T, in case that is required. The mean of this will be the mean time W spent by a customer in the system. This and other moments may be found directly from $L_T(s)$ using the moment generating properties of Laplace Transform. Using this approach, the following W may be obtained.

$$W = \frac{\overline{X^*}}{1 - \lambda\overline{X} + \lambda\overline{X^*}} + \frac{\lambda\overline{X^2}}{2(1 - \lambda\overline{X})} + \frac{\lambda(\overline{X^{*2}} - \overline{X^2})}{2(1 - \lambda\overline{X} + \lambda\overline{X^*})} \tag{4.21}$$

The overall mean service time will be $[(1-p_0)E\{X\}+p_0E\{X^*\}]$ taking into account the fact that the first customer in the busy period (probability p_0) will encounter a mean service time of $E\{X^*\}$, whereas the other customers (probability $1-p_0$) will have a mean service time of $E\{X\}$. Using this, the mean queueing delay W_q may be found to be

$$W_q = \frac{\lambda\overline{X^2}}{2(1 - \lambda\overline{X})} + \frac{\lambda(\overline{X^{*2}} - \overline{X^2})}{2(1 - \lambda\overline{X} + \lambda\overline{X^*})} \tag{4.22}$$

Knowing W and W_q, the mean N number in the system and the mean number N_q waiting in the queue may be found using Little's Result. It should be noted that even though the probability density function (actually the L.T. of the probability density function) result of Eq. (4.20) holds only for the FCFS queue, the mean results of Eqs. (4.21) and (4.22) and the results for N and N_q will hold for any service discipline, i.e. FCFS, LCFS or SIRO.

The above results may also be obtained by using a residual life based analytical approach. This is being left as an exercise for the user (see

Problem 1). Another typical way of describing a queue like the one considered in this section is to state that the first service in a busy period requires an additional Δ seconds of service (Δ random with its own moments and distribution) over and above the normal service time X. The analysis of such a system may be done in the same manner as above.

4.4 $M^{[X]}/G/1$ Queue - Single Server Queue with Batch Arrivals

We consider here a single server queue with batch arrivals and with a general service time distribution for the individual jobs within the batch. The following notation is used in the analysis. The arrival process of the batches is considered to be Poisson with average rate λ. A batch will contain r individual jobs ($1 \leq r \leq \infty$) where r is a discrete random variable with distribution β_r and generating function $\beta(z)$.

$$\beta(z) = \sum_{r=1}^{\infty} \beta_r z^r$$

The mean batch size $E\{r\}$ will then be given by

$$E\{r\} = \bar{r} = \beta'(1) = \sum_{r=1}^{\infty} r\beta_r$$

Note that it is also possible to model these using batch sizes that range from 0 to ∞. This would merely change the form of the expressions derived in this section but the analytical approach will remain the same.

The service times of the individual jobs are considered to be generally distributed. This distribution is assumed to have a given probability density function $b(t)$ and its corresponding L.T. is $L_B(z)$. (As usual, the service times are assumed to be i.i.d random variables.) In this case, it is also possible to define and use a *batch service time* as the total service time required by all the r jobs in a batch – the L.T. of the probability density function of this (given r) will be $[L_B(z)]^r$.

Consider an arbitrary interval of length t in this queueing system. Let $N(t)$ be the number of customer arrivals in time interval t and $\gamma(t)$ be the number of batches arriving in this time interval. Then, we have

$$P\{\gamma(t)=n\}=\frac{(\lambda t)^n e^{-\lambda t}}{n!} \tag{4.23}$$

Let the batch sizes be α_1, α_2, α_3, ... α_γ where $\alpha_i \perp \alpha_j$ for $i \neq j$ and $\{\alpha_i\}$ are independent, identically distributed (i.i.d.) random variables. We then get that

$$N(t)=N=\alpha_1+\alpha_2+\ldots\ldots+\alpha_\gamma \quad \text{(i.e. sum of i.i.d. random variables)}$$

$$E\{z^N \mid \gamma\}=\left[E\{z^\alpha\}\right]^\gamma=\beta^\gamma(z)$$

$$E\{z^N\}=\sum_{\gamma=0}^{\infty}\beta^\gamma(z)\frac{(\lambda t)^\gamma}{\gamma!}e^{-\lambda t}=e^{-\lambda t[1-\beta(z)]} \tag{4.24}$$

Note that $E\{z^N\}$ given by Eq. (4.24) is the generating function of the number of arrivals in an arbitrary interval of length t. If we now consider the time interval t to be actually a service time, we can get the generating function $A(z)$ of the number of arrivals within a service time to be

$$A(z)=\int_{t=0}^{\infty}e^{-\lambda t[1-\beta(z)]}b(t)dt=L_B(\lambda-\lambda\beta(z)) \tag{4.25}$$

If we imbed our Markov Chain once again at the customer departure instants then using the $A(z)$ given above, we can analyse this Markov Chain in the same way as for the normal M/G/1 queue. Let $Q(z)$ be the generating function of the number left behind in the system by a departing customer. Substituting the above value of $A(z)$from Eq. (4.25) in the earlier M/G/1 result (i.e. the P-K Transform equation of Eq. (3.14)), we will obtain $Q(z)$ as

$$Q(z)=\frac{(1-\rho)(1-z)L_B(\lambda-\lambda\beta(z))}{L_B(\lambda-\lambda\beta(z)-z)} \tag{4.26}$$

where $\rho=A'(1)=\lambda\overline{\beta}\overline{X}=\lambda\beta'(1)\overline{X}$

Unfortunately, this does not lead to results which are very usable [Kle75], The problem with Eq. (4.26) is that the state distribution given by $Q(z)$ is only valid at the customer departure instants. Since the queue has batch arrivals, positive state transitions can be more than +1 and hence Kleinrock's result cannot be applied - i.e., we cannot claim that this same distribution will hold for the arrival instants as well! PASTA will also not be applicable as the arrival process of the individual jobs is also no longer Poisson in nature.

To get more insight into the performance of the $M^{[X]}/G/1$ queue, we would need to approach the analysis of this queue in a different manner. This is discussed next.

4.4.1 An Alternate Approach to the Analysis of the $M^{[X]}/G/1$ Queue

In this approach, let us view the batch arrivals themselves as the service requests with their service times being the sum of the time taken to service all the customers within the batch. Let $b^*(t)$ be the probability density function of this *batch service time* X^* with L.T. $L_{B^*}(s)$. We can therefore see that

$$L_{B^*}(s) = \sum_{r=1}^{\infty} \beta_r \left(L_B(s)\right)^r = \beta(L_B(s)) \tag{4.27}$$

with

$$\overline{X^*} = -\left.\frac{dL_{B^*}(s)}{ds}\right|_{s=0} = \left[-L_B'(s)\beta'(L_B(s))\right]_{s=0} = \bar{r}\bar{X} = \bar{X}\beta'(1) \tag{4.28}$$

$$\begin{aligned}\overline{X^{*2}} = \left.\frac{d^2 L_{B^*}(s)}{ds^2}\right|_{s=0} &= \left[L_B''(s)\beta'(L_B(s)) + \{L_B'(s)\}^2 \beta''(L_B(s))\right]_{s=0} \\ &= \overline{X^2}\bar{r} + (\bar{X})^2[\overline{r^2} - \bar{r}]\end{aligned} \tag{4.29}$$

Let $\chi(z)=L_{B^*}(\lambda-\lambda z)=\beta(L_B(\lambda-\lambda z))$ be the generating function of the number of batch arrivals within a batch service time. Since we are considering the entire batch as a service request, we can imbed our Markov Chain at the batch departure instants. Let $Q^*(z)$ be the generating function of the number of batches left behind by a departing batch. The advantage of working in this way is that since the number of batches in the system can go up or down by at most *±1*, this distribution will also hold for the customer arrival instants (using Kleinrock's Result). Moreover, since the batch arrival process is Poisson, this same distribution will also hold for time averages at equilibrium. By analogy with our earlier analysis, we can then identify that

$$A(z) \rightarrow \chi(z) = L_{B*}(\lambda - \lambda z) = \beta(L_B(\lambda - \lambda z)) \tag{4.30}$$

with $\rho = \lambda \bar{r} \bar{X}$ (4.31)

This leads to

$$Q^*(z) = \frac{(1-\rho)(1-z)[\beta(L_B(\lambda - \lambda z))]}{[\beta(L_B(\lambda - \lambda z))] - z} \tag{4.32}$$

Let W_{qb} be the mean waiting time in queue for service to start to a batch. We can obtain this either from direct residual life-time arguments or from *$Q^*(z)$* above to be -

$$W_{qb} = \frac{\lambda}{2(1-\rho)} \overline{X^{*2}} \tag{4.33}$$

where ρ is the overall traffic value given by Eq. (4.31) and

$$\overline{X^{*2}} = (\overline{X^2} - \bar{X}^2)\bar{r} + \overline{r^2}\bar{X}^2 = \sigma_X^2 \bar{r} + \left(\bar{r}^2 + \sigma_r^2\right)\bar{X}^2 \tag{4.34}$$

Alternatively, we can also write W_{qb} in the form

$$W_{qb} = \frac{\rho \bar{r} \bar{X}}{2(1-\rho)} \left[1 + \frac{C_X^2}{\bar{r}} + C_r^2 \right] \tag{4.35}$$

with $C_X^2 = \dfrac{\sigma_X^2}{\bar{X}^2}$ and $C_r^2 = \dfrac{\sigma_r^2}{\bar{r}^2}$ as the *squared coefficients of variation* (also called SQV) of X and r.

We now assume that within a batch, customers are served in a random order. We then need to find the probability γ_k that a customer is served in the

k^{th} order in a batch. The queueing delay for this customer will then be the batch queueing delay W_{qb} and the sum of k service times - the mean of this will then be $W_{qb}+kE\{X\}$. Knowing γ_k, the mean of this overall queueing delay may be found. This will then be the average queueing delay as seen by an arbitrary customer (within a batch) in this $M^{[X]}/G/1$ queue.

To find γ_k, we proceed as follows. Let X_i be the batch size of the ith batch and let $y_i(k)$ be the probability that there is a call in the ith batch which is served in the kth order. Then, we can see that

$$\begin{aligned} y_i(k) &= 1 \quad \text{for } X_i \geq k \\ &= 0 \quad \text{otherwise} \end{aligned}$$

Now consider a set of n batches (note that we will eventually take the limit as $n \to \infty$). For these, we will have

$$\gamma_k = \lim_{n\to\infty} \frac{\sum_{i=1}^{n} y_i(k)}{\sum_{i=1}^{n} X_i} = \frac{\text{Number of calls served in the } k^{th} \text{ order}}{\text{Total number of calls in } n \text{ batches}}$$

This may then be simplified as follows

$$\gamma_k = \lim_{n\to\infty} \frac{\frac{1}{n}\sum_{i=1}^{n} y_i(k)}{\frac{1}{n}\sum_{i=1}^{n} X_i} = \frac{P\{X_i \geq k\}}{E\{X_i\}} = \frac{1}{\bar{r}}\sum_{i=k}^{\infty} \beta_i \tag{4.36}$$

Let W_2 be the random queueing delay for a call given that service to its batch has started. Assume that the random variable W_2 has probability density function $f_{W2}(t)$ with L.T. $L_{W2}(s)$. Then, we have

$$L_{W2}(s) = \sum_{k=1}^{\infty} [L_B(s)]^{k-1} \gamma_k = \frac{1}{\bar{r}} \sum_{k=1}^{\infty} [L_B(s)]^{k-1} (\beta_k + \beta_{k+1} + \beta_{k+2} + \ldots\ldots\infty)$$

This may be simplified to get

$$L_{W2}(s) = \frac{1}{\bar{r}} \sum_{i=1}^{\infty} \beta_i \frac{1-[L_B(s)]^i}{1-L_B(s)} = \frac{[1-\beta(L_B(s))]}{\bar{r}(1-L_B(s))} \tag{4.37}$$

Using the moment generating property of the L.T., we get

$$W_2 = L_{W2}(s)\big|_{s=1} = \frac{\bar{X}[\overline{r^2} - \bar{r}]}{2\bar{r}} = \bar{X}\left[\frac{\bar{r}}{2}(1+C_r^2) - \frac{1}{2}\right] \tag{4.38}$$

Therefore the overall mean queueing delay W_q for a job (in a batch) in an $M^{[X]}/G/1$ queue will be

$$W_q = W_{qb} + W_2 = \frac{\rho \bar{r} \bar{X}}{2(1-\rho)}\left[1 + \frac{C_X^2}{\bar{r}} + C_r^2\right] + \frac{\bar{X}}{2}\left[\bar{r}(1+C_r^2) - 1\right] \tag{4.39}$$

4.5 Single Server M/G/1 Priority Queues

Consider a single server M/G/1 queue where the customers have P different priority levels, *1* to P. We assume that customers of priority class 1 have the lowest priority whereas customers of priority class P have the highest priority. It is possible to conceptually think of the queue as being organised sequentially in the buffer of the single server queue as shown in Figure 4.2.

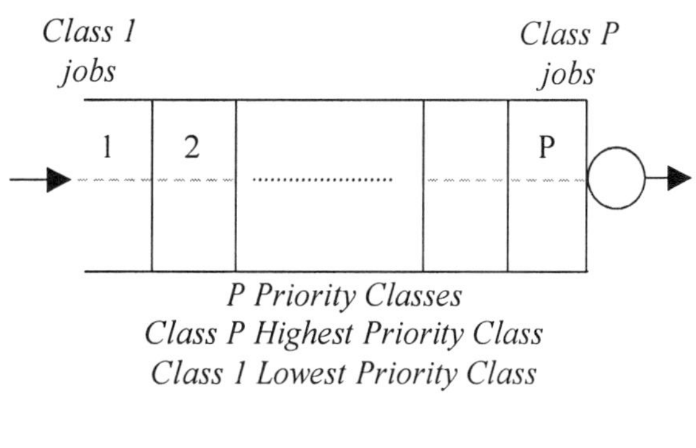

Figure 4.2. M/G/1 Queue with Head of Line Priority

The queueing arrangement illustrated in Figure 4.2 is often referred to as Head-of-Line (HOL) priority. This emphasises the fact that if the queue is FCFS in nature, then an arrival of priority class k joins the queue after the last class k arrival that is presently queued. Whenever the server finishes service to a job/customer, it picks up the next customer to be served from the front of the queue, whatever its present state may be. Additional rules are still required to handle the situation when a customer of a higher priority class (say j), arrives when a customer of relatively lower priority (say k, $k<j$) is in service. When this happens, we can have several choices on the approach to be followed for handling the presently ongoing service and the service to the newly arriving customer of higher priority class. The common approaches for handling such a situation are the following.

[A] Non-preemptive Priority

Consider an arrival of priority j to the system which finds a customer of priority k currently being served where $j>k$. If the *Non-preemptive Priority* discipline is followed, the new customer joins the queue of its own priority class without interrupting the ongoing service of the customer of class k even though the newly arriving customer has a higher priority level. Once the current service gets completed, the server examines the queue to start service to the customer at the front of the highest priority class with a non-empty queue.

[B] Preemptive Resume Priority

Consider an arrival of priority j to the system which find a customer of priority k currently being served where $j>k$. If the *Preemptive Resume Priority* discipline is followed then the new arrival immediately *preempts* the service of the lower priority customer currently being served. Service to the preempted customer *resumes from the point of interruption* when that priority class is the highest priority class with a non-empty queue once again. Note that in this case, the work done on the preempted customer is conserved as service resumes from the point of interruption, i.e. work done is not lost because of the preemption. (Note that work will also not be lost for the *Non-preemptive Priority* case.)

[C] Preemptive Non-Resume Priority

Consider an arrival of priority j to the system which find a customer of priority k currently being served where $j>k$. In this case also the lower priority customer currently being served gets preempted by the new higher priority arrival. However, unlike the preemptive resume strategy, when service resumes for the preempted customer it starts once again from the beginning, i.e. the service that has already been provided when the customer was preempted is lost. This priority approach is non-work conserving in nature. Its analysis is difficult and will not be attempted here.

We introduce below the common notation that we will use for studying this type of queues in this section.

(*i*) Average arrival rate for priority class $i = \lambda_i \quad i=1,\ldots\ldots, P$

We assume that the arrival process of each class is individually Poisson in nature and that the arrival processes of different classes are independent of each other. The total arrival process will also be also be Poisson as the sum of independent Poisson processes is also a Poisson process.

(*ii*) Service time for priority class i has mean $\overline{X}_i$ with cumulative distribution function $B_i(t)$ and probability density function $b_i(t)$ with L.T. $L_{Bi}(s)$, for $i=1,\ldots\ldots,P$ The service times for the different priority classes are assumed to be independent of each other.

(*iii*) Traffic of priority class $i = \rho_i = \lambda_i \overline{X}_i \qquad i=1,\ldots\ldots, P$

(*iv*) Total Arrival Rate = $\lambda = \sum_{i=1}^{P} \lambda_i$

(*v*) Average Overall Service Time = $\overline{X} = \sum_{i=1}^{P} \frac{\lambda_i}{\lambda} \overline{X}_i$

(*vi*) Total Traffic = $\rho = \lambda \overline{X} = \sum_{i=1}^{P} \rho_i$

In the following, we analyse the steady state behaviour of simple, single-server, priority queues with infinite buffer capacities. One should also note that for priority queues, it is possible for the queue to become unstable for lower priority traffic while still being stable for the higher priorities. As an example, consider the preemptive resume, two-priority case with class 1 of low priority and class 2 of high priority, where this combined traffic is offered to a M/G/1 queue. If $\rho_2<1$, then the queue will be stable with respect to class 2 priority traffic and equilibrium conditions will exist for this class. However, if at the same time, $(\rho_1+\rho_2)>1$, then the queue will be unstable if one takes into account the low priority traffic as well. To keep our analysis simple, we will normally assume that the traffic $\{\rho_i\}$ are such that the queue is stable for all priority classes. This implies that the condition

$$\rho = \sum_{i=1}^{P} \rho_i < 1$$

will hold since the queue is single-server in nature.

4.5.1 Residual Life Analysis for the Non-Preemptive Priority M/G/1 Queue

We assume a queue with P priority classes of users with class 1 with the lowest priority and class P with the highest priority. Let N_{qk} be the mean number waiting in queue for the priority class k and let W_{qk} be the corresponding mean waiting time in queue. Note that Little's result will still hold for this and that therefore $N_{qk}=\lambda_k W_{qk}$. Consider a particular arrival of class k which enters the queue. Let R be the mean residual time that it will see to finish service to whichever customer is currently being served. Note that R will not depend on the priority class k of the arriving customer as it will be the same for an arrival from any priority class. This will be true because of the non-preemptive priority discipline followed by this queue. Using the typical graphical approach followed for the residual life based analysis earlier, we get the mean residual time R (for arrival of any priority class) to be

$$R = \frac{1}{2}\sum_{i=1}^{P} \lambda_i \overline{X_i^2} \tag{4.40}$$

where $\overline{X_i^2}$ is the second moment of the service time for the ith priority class.

Consider an arrival of class P (the highest priority class), to this queue. For this arrival, we can write

$$W_{qP} = R + \overline{X}_P N_{qP}$$

Using Little's result for this priority class (i.e. $N_{qP}=\lambda_P W_{qP}$), we get

$$W_{qP} = \frac{R}{1-\rho_P} \tag{4.41}$$

Similarly, for an arrival of class *(P-1),* we can write

$$W_{q(P-1)} = R + \overline{X}_P N_{qP} + \overline{X}_{P-1} N_{q(P-1)} + \overline{X}_P \lambda_P W_{q(P-1)} \tag{4.42}$$

It is worthwhile to examine the individual delay terms in the R.H.S. of Eq. (4.42). The first term R gives the mean residual service time to finish service to the customer (if any) currently being served (when the customer of interest of class *P-1* arrives). The term $\overline{X}_P N_{qP}$ is the mean time taken to serve all the waiting customers of class P who are presently waiting in the queue. Similarly, $\overline{X}_{P-1} N_{q(P-1)}$ is the mean time taken to serve all those class *P-1* customers, who are presently waiting in the queue. Finally the term $\overline{X}_P \lambda_P W_{q(P-1)}$ is the mean time taken to serve all the customers of class P who arrive while the class *P-1* customer of interest is waiting in the queue. Simplifying Eq. (4.42) using $N_{qP}=\lambda_P W_{qP}$ and $N_{q(P-1)}=\lambda_{P-1} W_{q(P-1)}$, we get

$$W_{q(P-1)}(1-\rho_P-\rho_{P-1}) = W_{qP} \tag{4.43}$$

Using Eq. (4.41), this may be simplified to get the mean queueing delay for class (P-*1*) customers as

$$W_{q(P-1)} = \frac{R}{(1-\rho_P)(1-\rho_P-\rho_{P-1})} \tag{4.44}$$

For class *(P-2)*, we can similarly show that

$$\begin{aligned} W_{q(P-1)} &= R + \overline{X}_P N_{qP} + \overline{X}_{P-1} N_{q(P-1)} + \overline{X}_{P-2} N_{q(P-2)} \\ &\quad + \overline{X}_P \lambda_P W_{q(P-2)} + \overline{X}_{P-1} \lambda_{P-1} W_{q(P-2)} \end{aligned} \tag{4.45}$$

which on simplification leads to

$$W_{q(P-2)} = \frac{R}{(1-\rho_P-\rho_{P-1})(1-\rho_P-\rho_{P-1}-\rho_{P-2})} \tag{4.46}$$

We can therefore derive the general result for the mean queueing delays for each of the *P* classes as

$$W_{qP} = \frac{R}{1-\rho_P} \qquad \text{for } i=P$$

$$W_{q(P-i)} = \frac{R}{(1-\sum_{j=0}^{i-1}\rho_{P-j})(1-\sum_{j=0}^{i}\rho_{P-j})} \qquad \text{for } i=1,\ldots\ldots,(P-1) \tag{4.47}$$

Once the queueing delay W_{qi} $i=1,\ldots,P$ is found, the mean time spent in system by a customer of priority class *i* may be found using

$$W_i = W_{qi} + \overline{X}_i \tag{4.48}$$

Using Little's result, the mean number N_i of class *i* jobs in the system and the mean number N_{qi} of class *i* jobs waiting in queue may then be found from W_i and W_{qi}, respectively, for $i=1,\ldots.,P$

4.5.2 Residual Life Analysis for the Preemptive Resume Priority M/G/1 Queue

In this case, the service to customers of priority *i* gets interrupted if arrivals occur from priority classes *i+1* or higher. The interrupted service resumes from the point of interruption when all customers of priority classes *i+1* or higher in the queue have been served. As in the non-preemptive case, there is no loss of work. It should also be noted that the service to customers of priority class *k* is completely unaffected by the service demands of customers of lower priority classes, i.e. priority classes *1, ..., (k-1)*. This point will be useful in the subsequent analysis based on residual lifetimes.

In this case, one cannot meaningfully define a mean queueing delay (i.e. time spent waiting in queue prior to service) for priority classes *1, ..., (P-1)*, i.e. for priority classes other than the highest priority class. This is because even after service to such a customer starts, the service may be preempted and the customer will be forced back into the queue to wait once again. However, the mean total time W_i spent in system by a customer of priority class i may still be defined in the usual fashion as the mean time taken for an arriving customer to leave the system.

In contrast to the non-preemptive case, a customer arrival of class k will see a mean residual service time R_k which will depend on its own class. This is because this customer will see the residual service time only from on-going service of class k or higher. Following the usual graphical derivation approach, we can get

$$R_k = \sum_{i=k}^{P} \frac{1}{2}\lambda_i \overline{X_i^2} \qquad \text{for } k=1, 2, \ldots\ldots, P \tag{4.49}$$

Consider an arrival of the highest priority class P. In this case, we can define a queueing delay W_{qP} in the usual fashion and can write -

$$W_{qP} = R_P + \overline{X}_P N_{qP} \quad \text{and} \quad N_{qP} = \lambda_P W_{qP}$$

Therefore

$$W_{qP} = \frac{R_P}{1-\rho_P} \tag{4.50}$$

and hence

$$W_P = W_{qP} + \overline{X}_P = \frac{\overline{X}_P(1-\rho_P) + R_P}{(1-\rho_P)} \tag{4.51}$$

For an arrival of class $(P\text{-}1)$, we can write

$$W_{P-1} = \overline{X}_{P-1} + \frac{R_{P-1}}{1-\rho_P-\rho_{P-1}} + \overline{X}_P \lambda_P W_{P-1} \tag{4.52}$$

In the R.H.S. of Eq. (4.52), the component $\overline{X}_{P-1}$ is obvious as the mean time taken to actually serve the customer of interest of class $P\text{-}1$. The term

$\frac{R_{P-1}}{1-\rho_P-\rho_{P-1}}$ is analogous to writing $W_q = \frac{R}{1-\rho}$ as the mean time to serve those customers of class *P* and *P-1* who are already in system when the customer of interest (of class *P-1*) arrives. Note that this term may be argued from the observation that we can effectively ignore the presence of priority classes *1,....,(P-2)* for these customers because of the nature of the preepmptive resume priority operation. We can interpret $\rho_P + \rho_{P-1}$ as ρ and R_{P-1} as *R* in that expression. The mean residual service time R_{P-1} observed by an arrival of class *P-1* will depend on the traffic of classes *P* and *P-1* but will be unaffected by traffic of lower priority classes. (Note that unlike the non-preemptive case of Sec. 4.5.1, the mean residual service time is class dependent because of the preemptive nature of the priority mechanism.) The expression for this is given in Eq. (4.49). The term $\overline{X}_P \lambda_P W_{P-1}$ corresponds to the mean time taken to serve all those customers of class *P* who arrive while the customer of interest (of class *P-1*) is in the system. Since these class *P* customers will also get served before the service to the customer of interest concludes, this also needs to be included in Eq. (4.52). Simplifying the above, we get

$$W_{P-1} = \frac{\overline{X}_{P-1}(1-\rho_P-\rho_{P-1})+R_{P-1}}{(1-\rho_P)(1-\rho_P-\rho_{P-1})} \tag{4.53}$$

Similarly, for an arrival of class *(P-2)*, we will have

$$W_{P-2} = \frac{\overline{X}_{P-2}(1-\rho_P-\rho_{P-1}-\rho_{P-2})+R_{P-2}}{(1-\rho_P-\rho_{P-1})(1-\rho_P-\rho_{P-1}-\rho_{P-2})} \tag{4.54}$$

The mean time spent in system for other classes may be similarly found. For example, for the lowest priority customers of class 1, we will get

$$W_1 = \frac{\overline{X}_1(1-\rho_P-......-\rho_1)+R_1}{(1-\rho_P-......-\rho_2)(1-\rho_P-......-\rho_1)} \tag{4.55}$$

The term $(1-\rho_P -....-\rho_i)$ in the denominator of the mean delay W_i for a customer of priority class *i* in Eq. (4.50) and Eqs. (4.53)-(4.55) is important. Its presence shows that the mean delay for priority class *i* for this priority mechanism will stay bounded only if $(\rho_P +....+\rho_i)<1$ as had been stated earlier. Depending on the traffic values, the mean delay may become unbounded for a lower priority class even though it is bounded for higher classes.

The analysis for the priority queues given above has been done using a residual life approach. As discussed earlier, this approach can only give mean results for the performance parameters like mean number in the system, mean number waiting in queue, waiting time in queue and mean time spent in system. If actual distributions are desired, then we need to follow an imbedded Markov Chain approach with the imbedded points appropriately defined, i.e. as the departure points of the corresponding customers. This analysis can be done but tends to be quite complex. We illustrate this approach below for the *Two Priority Preemptive Resume* M/G/1 queue. This approach may be extended [Hay84] for more priority classes but the approach and the expressions become considerably more complex as the number of priority classes increase.

4.5.3 Imbedded Markov Chain Analysis for the Preemptive Resume Priority M/G/1 Queue

Consider the preemptive resume M/G/1 queue with just two classes of customers - high priority customers of class 2 and low priority customers of class 1 - with the notation as used in the earlier residual life based analysis. We analyse the queue separately for the two classes of customers.

Queue for Class 2 Customers

Note that for class 2 customers, the system would behave just like a simple M/G/1 queue with average arrival rate λ_2. Given that a class 2 service time has the probability density function $b_2(t)$ with $L_{B2}(s)$ as the corresponding Laplace Transform, the system may then be analysed for this priority class just like the simple M/G/1 queue of Chapter 3 using the approach of Section 3.2. This is because the presence of class 1 customers will not affect the performance of the queue for class 2 customers in any way. Specifically, the generating function $P_2(z)$ for the number of class 2 customers in the system may then be obtained as for Eq. (3.14) in Section 3.2.

$$P_2(z) = \frac{(1-\rho_2)(1-z)L_{B2}(\lambda_2 - \lambda_2 z)}{L_{B2}(\lambda_2 - \lambda_2 z) - z} \tag{4.56}$$

As done for the case of the simple M/G/1 queue in Sec 3.3, we can do a busy period analysis for this priority queue as well, considering only the class 2 customers. This will be the busy period only for class 2 customers in the queue, i.e. when the server is engaged in serving the higher priority class 2 customer. Let $f_{BP2}(t)$ be the probability density function of this class 2 busy period with its L.T. given by $L_{BP2}(s)$. This may be obtained by applying the results of Section 3.3. Specifically, using Eq. (3.19) we can find

$$L_{BP2}(s) = L_{B2}(s + \lambda_2 - \lambda_2 L_{BP2}(s)) \tag{4.57}$$

The mean number N_2 of class 2 users in the system may be found directly from Eq. (4.56) using the moment generating property of the generating function. Application of Little's result to N_2 will give the mean of the total time W_2 spent in the system by a class 2 arrival. We can then find W_{q2} as this will be the difference of W_2 and the mean service time of a class 2 job. Little's result may once again be applied to W_{q2} to find the mean number of class 2 jobs queued for service.

Queue for Class 1 Customers

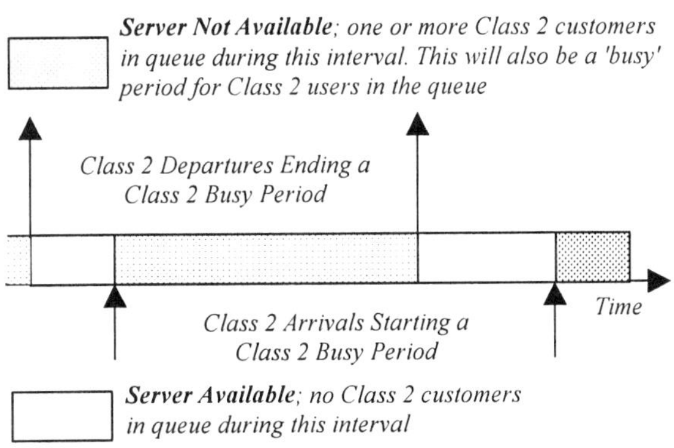

Figure 4.3. Server Available/Unavailable Intervals for Class 1 Customers

The lower priority queue needs separate analysis as the service to lower priority customers can get preempted by the service demands from the class 2 users. From the point of view of class 1 users, the time axis of the queue may be divided into two kinds of intervals - *server available* and *server unavailable* intervals. These are illustrated in Figure 4.3. It is easy to see that the *Server Available* (for class 1) intervals shown above will be exponentially distributed with mean $1/\lambda_2$ (i.e. with probability density function $\lambda_2 exp(-\lambda_2 t)$ for $t \geq 0$). This is because these correspond to the *Idle*

Periods of the class 2 queue. Considered from the point of view of the class 1 users, there is another important difference between this priority queue and a normal M/G/1 queue. In a simple M/G/1 queue, a new arrival, which finds the system empty, will start service immediately. In the priority queue being discussed here, a new class 1 arrival, which comes at a time when there is no other class 1 user in the system, may not start its service immediately. Specifically, a new class 1 arrival may be forced to wait in queue (even when there are no other class 1 customers in the system) if it comes during the *Server Unavailable* periods when the queue is actually serving class 2 customers. The class 1 queue is analysed by selecting the imbedded Markov points at the *departure instants of the class 1 customers*. As before (for the simple M/G/1 queue), we define n_i to be the number in the system as seen by the ith departing customer. In order to properly define a_{i+1} and $\hat{a}_{i+1}$, we need to define the "service time" properly for this class of customers. Note that because of potential interruptions from class 2 customers, the service time of a class 1 customer may not be a continuous interval. In this case, we define the *"service time" of a class 1 user* to be the sum of the actual service duration and the time during which the server is unavailable (during this service) because of preemption and the consequent serving of class 2 customers. With this definition of "*service time*", we can then define

a_{i+1} = Number of class 1 arrivals during the "service time" of the $(i+1)^{\text{th}}$ class 1 user

$\hat{a}_{i+1}$ = Number of class 1 arrivals during the time of waiting for the server and the "service time" of the $(i+1)^{\text{th}}$ class 1 customer.

We can then write, as before, that

$$n_{i+1} = n_i - U(n_i) + \hat{a}_{i+1} + (a_{i+1} - \hat{a}_{i+1})U(n_i) \tag{4.58}$$

or equivalently

$$\begin{aligned} n_{i+1} &= n_i - 1 + a_{i+1} && \textit{for } n_i \geq 1 \\ &= \hat{a}_{i+1} && \textit{for } n_i = 0 \end{aligned} \tag{4.59}$$

Note that the case for $\hat{a}_{i+1}$ arises only when $n_i = 0$. This is because of the following reasons.

(*i*) a_{i+1} is for the case where the queue has at least one class 1 user (already waiting in the queue) for whom service can begin immediately after the service completion of the previous class 1 user.

(*ii*) $\hat{a}_{i+1}$ is for the case where the class 1 queue is empty when the previous class 1 service is completed. Service to the new class 1 arrival may get delayed if it comes at a point in time when the server is not available for class 1 - i.e. server is serving class 2. If this happens then service to the class 1 arrival can start only after the class 2 queue becomes empty - this is what has been referred to as the "time of waiting for the server" in the definition given above for $\hat{a}_{i+1}$.

Using Eqs. (4.58) or (4.59), we can show as before, that at steady state, the generating function $P_1(z)$ of the number of class 1 users in queue at the class 1 departure instants, will be given by

$$P_1(z) = \frac{p_0[A(z) - z\hat{A}(z)]}{A(z) - z} \tag{4.60}$$

where

$A(z)$ = generating function of a, i.e. a_{i+1} at steady state (found later)

$\hat{A}(z)$ = generating function of $\hat{a}$, i.e. $\hat{a}_{i+1}$ at steady state (found later)

The probability p_0 of the system being empty of class 1 customers may be found from Eq. (4.60) by using $P_1(z)=1$ for $z=1$. This yields

$$p_0 = \frac{1-\bar{a}}{1-\bar{a}+\bar{\hat{a}}} \tag{4.61}$$

with $\bar{a}$ and $\bar{\hat{a}}$ as the means of a and $\hat{a}$, respectively. These means may be obtained using the normalisation conditions $P_1(1) = A(1) = \hat{A}(1) = 1$ and the moment generating properties $A'(1) = \bar{a}$ and $\hat{A}'(1) = \bar{\hat{a}}$. Therefore the complete solution would require that we obtain $A(z)$ and $\hat{A}(z)$ as given next. Incidentally, the fact that $A(z)$ and $\hat{A}(z)$ are different may also be viewed as an example of *exceptional first service*.

Obtaining A(z)

For this we need to consider the case where the departure of a class 1 customer leaves the class 1 queue non-empty. This has been illustrated in Figure 4.4 where the nth class 1 customer leaves at time t_0. At that instant, there are one or more class 1 users still in the queue and service to the $(n+1)^{th}$ class 1 customer starts immediately. Figure 4.4 shows that, after a number of service interruptions (caused because of class 2 arrivals), the $(n+1)^{th}$ class 1 customer finally leaves the system at time t_n.

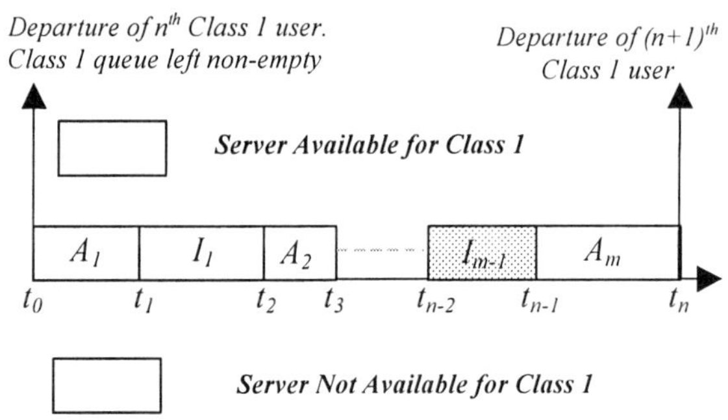

Figure 4.4. Class 1 Departure Leaving Class 1 Queue Non-Empty

In Figure 4.4, we have shown the service of the $(n+1)^{th}$ class 1 customer to be interrupted for the first time by a class 2 arrival at time t_1 and the corresponding busy period is (t_1, t_2) during which the server is not available to the $(n+1)^{th}$ class 1 customer whose service had started. This service resumes at t_2 and continues until the next interruption at time t_3. This process continues until the service to this class 1 customer finally concludes at time t_n. Note that in the above figure, the intervals $A_1, \ldots, A_m$ correspond to the server available (for class 1) intervals and the intervals $I_1, \ldots, I_{m-1}$ correspond to the intervals when the server is not available. As mentioned before, the intervals $\{A_i\}$ are independent, identically distributed (i.i.d.) random variables which are exponentially distributed with mean λ_2^{-1}. Similarly, the interruption intervals $\{I_i\}$ are also i.i.d. random variables with the L.T. of their probability density function obtained by solving Eq. (4.57) for $L_{BP2}(s)$. This may then be inverted to obtain the desired probability density function $f_{BP2}(s)$. We use the following notation, as appropriate, in the following derivation.

T = total *service time* for this class 1 user (including interruptions)
x_1 = actual service time (i.e. the real service time) whose probability density function is $b_1(t)$ with L.T. $L_{B1}(s)$
I_i = duration of the ith interruption
n = number of service interruptions

Then, it may be easily seen that T will consist of the real service time of the class 1 job and the interruptions encountered from class 2 arrivals during its service. this gives

$$T = x_1 + \sum_{i=1}^{n} I_i \tag{4.62}$$

Taking the mean of Eq. (.62), we will also get that

$$\bar{T} = \bar{X}_1 + \lambda_2 \bar{X}_1 \bar{I} = \bar{X}_1 + \lambda_2 \bar{X}_1 \overline{BP_2} \tag{4.63}$$

where $\overline{BP_2}$ is the mean busy period for the class 2 queue. It can also be easily shown that if the duration of the class 1 service is actually x_1 (i.e. the real service time), then the probability of n interruptions in its service will be given by the Poisson distribution $\left[[\exp(-\lambda_2 x_1)] \frac{(\lambda_2 x_1)^n}{n!} \right]$. This would follow from the fact that the lengths of the *server available* intervals are i.i.d. with an exponential distribution of mean λ_2^{-1} and that the sum of these add up to x_1. If t is the total service time of a class 1 user in the queue, whose actual service time requirement is u, and if n is the number of service interruptions encountered during its service, then (a) these service interruption intervals are i.i.d. random variables and (b) their total time is (t-u). The probability density function of one such service interruption will be $f_{BP2}(t)$ with L.T. $L_{BP2}(s)$. The sum of n such random variables will have the probability density function $f_{BP2}^{(n)}(t)$ which is the n-fold convolution of $f_{BP2}(t)$; the corresponding L.T. will be $[L_{BP2}(s)]^n$. Therefore , we have

$$P\{t<T\leq t+dt \mid x_1=u,\ n \text{ interruptions}\} = f_{BP2}^{(n)}(t-u)dt$$

implying that

$$P\{t<T\leq t+dt \mid x_1=u\}=\sum_{n=0}^{\infty}[\exp(-\lambda_2 u)]\frac{(\lambda_2 u)^n}{n!}f_{BP2}^{(n)}(t-u)dt$$

If $f_T(t)$ is the probability density function of T with L.T. $L_T(s)$, then we have

$$\begin{aligned} f_T(t)dt &= P\{t<T\leq t+dt\} \\ &= \int_{u=0}^{\infty}\sum_{n=0}^{\infty}[\exp(-\lambda_2 u)]\frac{(\lambda_2 u)^n}{n!}f_{BP2}^{(n)}(t-u)b_1(u)dudt \end{aligned}$$

Taking the Laplace Transform of $f_T(t)$, we obtain

$$\begin{aligned} L_T(s) &= \int_{t=0}^{\infty} e^{-st} f_T(t)dt \\ &= \int_{t=0}^{\infty} e^{-st}\left[\int_{u=0}^{\infty}\sum_{n=0}^{\infty}[\exp(-\lambda_2 u)]f_{BP2}^{(n)}(t-u)b_1(u)du\right]dt \end{aligned}$$

Changing the order of integration and identifying $L_{BP2}(s)$ as the Laplace Transform of $f_{BP2}(t)$, we get that

$$\begin{aligned} L_T(s) &= \int_{u=0}^{\infty} b_1(u)[\exp(-\lambda_2 u)]\left[\sum_{n=0}^{\infty}\frac{(\lambda_2 u)^n}{n!}(L_{BP2}(s))^n\right]e^{-su}du \\ &= \int_{u=0}^{\infty}[\exp(-u[s+\lambda_2-\lambda_2 L_{BP2}(s)])]b_1(u)du \qquad (4.64) \\ &= L_{B1}(s+\lambda_2-\lambda_2 L_{BP2}(s)) \end{aligned}$$

The L.T. $L_T(s)$ of the probability density function $f_T(t)$ of the random variable T may also be obtained by an alternate, somewhat easier, approach. For this, note that

$$E\{e^{-sT} \mid u, n\} = (L_{BP2}(s))^n e^{-su} \tag{4.65}$$

Removing the conditioning on n in Eq. (4.65), we get

$$\begin{aligned} E\{e^{-sT} \mid u\} &= \sum_{n=0}^{\infty} (L_{BP2}(s))^n e^{-su} \frac{(\lambda_2 u)^n}{n!} [\exp(-\lambda_2 u)] \\ &= \exp[-su - \lambda_2 u + \lambda_2 u L_{BP2}(s)] \\ &= \exp[-u(s + \lambda_2 - \lambda_2 L_{BP2}(s))] \end{aligned} \tag{4.66}$$

Finally, we remove the conditioning on u in Eq. (4.66)

$$\begin{aligned} L_T(s) &= \int_{u=0}^{\infty} [\exp - u(s + \lambda_2 - \lambda_2 L_{BP2}(s)] b_1(u) du \\ &= L_{B1}(s + \lambda_2 - \lambda_2 L_{BP2}(s)) \end{aligned}$$

to get the same result as in Eq. (4.64) once again.

Since $A(z)$ is the generating function of the number of class 1 customers arriving in the time interval T, we can then write

$$A(z) = L_T(\lambda_1 - \lambda_1 z) \tag{4.67}$$

To summarise, the overall procedure for getting $A(z)$ will therefore be

(a) Solve Eq. (4.57), $L_{BP2}(s) = L_{B2}(s + \lambda_2 - \lambda_2 L_{BP2}(s))$, for $L_{BP2}(s)$

(b) Get $L_T(s)$ from Eq. (4.64), $L_T(s) = L_{B1}(s + \lambda_2 - \lambda_2 L_{BP2}(s))$

(c) Get $A(z)$ from Eq. (4.67), $A(z) = L_T(\lambda_1 - \lambda_1 z)$

Obtaining $\hat{A}(z)$

For this we need to consider the case where the departure of a class 1 customer leaves the class 1 queue empty as illustrated in Figure 4.5. As shown in Figure 4.5, assume that the nth class 1 departure occurs at time $t=0$.

However, since in this case, there are no other class 1 users in the queue at that instant, the server will actually become idle at this point. (The fact that a class 1 user was being served at $t=0^-$ of course implies that there cannot be a class 2 user in the queue at that instant.) There will be a sequence of "server available for class 1" $\{A_i\ i=1,2,....\}$ and "server unavailable for class 1" $\{I_j\ j=1,2,,,,,)$ intervals after this as shown in the figure. Note that the next period when the server is busy serving a class 1 customer cannot begin until there is an actual class 1 arrival. Service to this newly arriving class 1 customer, i.e. the $(n+1)^{th}$ one, may or may not begin from the instant of its arrival to the queue. If this arrival happens at a time when the server is unavailable to serve class 1 customers (i.e. if it is busy serving some higher priority class 2 arrival which came after $t=0$), then service to the $(n+1)^{th}$ class 1 customer will not begin until the queue once again becomes empty of class 2 customers. However, if the $(n+1)^{th}$ class 1 customer does arrive during one of the "server available for class 1" interval (when the queue will actually be empty), then service to it will begin from its arrival instant itself. Once service to the $(n+1)^{th}$ class 1 customer begins, it will proceed, possibly with interruptions cause by pre-emption by class 2 customers, until its required service is over. When that happens, the $(n+1)^{th}$ class 1 service time will be completed and that customer will depart at time $t=t_n$ as shown in Figure 4.5.

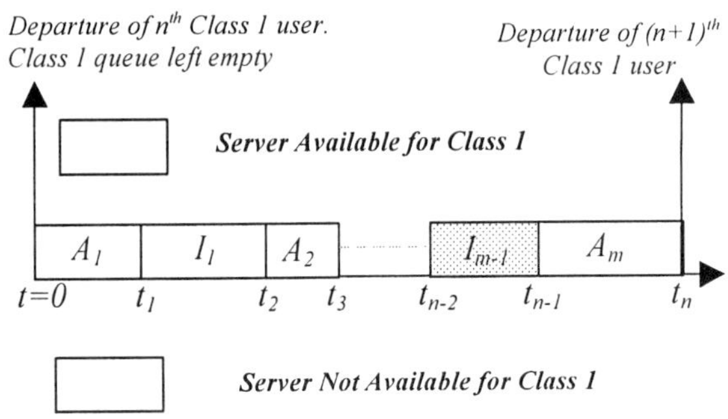

Figure 4.5. Class 1 Departure Leaving Class 1 Queue Empty

Let $t=\tau$ be the time instant (random variable, measured from $t=0$) when the next class 1 customer arrives - i.e. after the class 1 departure at $t=0$ which left the class 1 queue empty. Then, we can make the following observations -

(a) τ is exponentially distributed with mean λ_1^{-1} as the class 1 arrivals come from a Poisson process with mean arrival rate λ_1

(b) Arrival may fall either in an A_i interval or in an I_j interval. If it falls in A_i, then service starts immediately; if it falls in I_j, then service starts after I_j is over

(c) The $\{A_i\}$ intervals are i.i.d. with each of them having an exponential distribution with mean λ_2^{-1}

(d) The $\{I_j\}$ intervals are i.i.d. random variables and are also independent of the $\{A_i\}$ intervals. Note that the $\{I_j\}$ intervals are actually the busy periods for the class 2 customers in the queue. These will have the probability density functions as $f_{BP2}(t)$ with L.T. $L_{BP2}(s)$ obtainable by solving Eq. (4.57). The probability density function $f_{BP2}(t)$ may then be obtained by taking the inverse transform of $L_{BP2}(s)$.

Let $WQ1$ be the (random) time between the arrival of a class 1 user (to an empty class 1 queue) and its start of service. Let the probability density function of this be $f_{WQ1}(t)$ with L.T. $L_{WQ1}(s)$. Actually, what we need to find is the distribution of the time interval T^* for which an arrival, such as this $(n+1)^{th}$ one, stays in the system, given that the previous class 1 departure left the class 1 queue empty. Note that T^* will actually be the sum of T, as defined for finding $A(z)$ earlier, and $WQ1$ where the two random variables are independent of each other. It should be recalled that T, as defined for finding $A(z)$, is the time between the start of service to a class 1 customer and the completion of this service, including the time intervals when the service gets preempted by class 2 customers. We can therefore write that

$$T^* = T + WQ1 \qquad T \perp WQ1 \tag{4.68}$$

and therefore,

$$L_{T^*}(s) = L_T(s)L_{WQ1}(s) \tag{4.69}$$

Since $L_T(s)$ can be found using Eq. (4.64), we really just need to find $L_{WQ1}(s)$ to find the required $L_{T^*}(s)$. Once this is known, we can follow our usual approach to show that since the class 1 arrivals, come from a Poisson process with mean rate λ_1, we would have

$$\hat{A}(z) = L_{T^*}(\lambda_1 - \lambda_1 z) = L_T(\lambda_1 - \lambda_1 z)L_{WQ1}(s)(\lambda_1 - \lambda_1 z) \tag{4.70}$$

We therefore need to find the distribution of $WQ1$, the queueing delay to start service to a class 1 arrival when the previous class 1 departure left the class 1 queue empty. For this, we need to consider separately, the following two cases -

(a) *WQ1=0*, i.e. the class 1 arrival happens in some interval A_i when the server is available. In this case, the arrival starts service from the instant of arrival. Note that in this case the arrival can start service immediately as it enters the queue at a time when the server is not engaged in serving other customers
(b) *WQ1>0*, i.e. the class 1 arrival happens in some interval I_j when the server is not available. In this case, the arrival is queued to wait until the end of this I_j interval as that is the time when the server will once again become available for serving class 1 customers.

We consider first the case when *WQ1=0* where the $(n+1)^{th}$ class 1 arrival happen at time τ. For this, we can show that

$$P\{WQ1=0\}=\sum_{n=0}^{\infty}P\left\{\sum_{i=0}^{n}(A_i+I_i)<\tau\le\sum_{i=0}^{n}(A_i+I_i)+A_{n+1}\right\} \tag{4.71}$$

Note that for notational convenience, we have set $A_0=F_0=0$ in Eq. (4.71). This allows a more compact form of the equation without any loss of generality. Note that τ is an exponentially distributed ranom variable. It will therefore be memory-less and its memory-less property may be utilized to show that

$$P\left\{\sum_{i=0}^{n}(A_i+I_i)<\tau\le\sum_{i=0}^{n}(A_i+I_i)+A_{n+1}\right\} \\ =\left[\prod_{i=0}^{n}P\{A_i<\tau\}P\{I_i<\tau\}\right]P\{A_{n+1}\ge\tau\} \tag{4.72}$$

The argument for the above may be given as follows. If τ does not fall in A_i then the probability that $\{\tau$ does not fall in $I_i\}$, i.e. P$\{\tau$ does not fall in $I_i\}$, will be P$\{I_i<\tau\}$. This will hold as one may consider τ as beginning from the start of I_i and use the fact that the inter-arrival times for a Poisson arrival process (i.e. the class 1 and class 2 arrival processes) will be exponentially distributed and hence memoryless. Similar arguments may be given for the probability terms P$\{A_i<\tau\}$ and P$\{A_{n+1}\ge\tau\}$ in Eq. (4.72). We can then use the typical properties of the Poisson arrival processes of class 1 and class 2 traffic to get

$$
\begin{aligned}
P\{A_i < \tau\} &= P\{Class\ 2\ arrival\ occurs\ before\ a\ Class\ 1\ arrival\} \\
&= \int_{t=0}^{\infty} [\exp(-\lambda_2 t)](\lambda_2 dt) \int_{\alpha=t}^{\infty} [\exp(-\lambda_1 \alpha)](\lambda_1 d\alpha) \\
&= \int_{t=0}^{\infty} \lambda_2 [\exp(-\lambda_2 t)][\exp(-\lambda_1 t)] dt \\
&= \frac{\lambda_2}{\lambda_1 + \lambda_2}
\end{aligned} \tag{4.73}
$$

$$
\begin{aligned}
P\{A_{n+1} \geq \tau\} &= P\{Class\ 1\ arrival\ occurs\ before\ a\ Class\ 2\ arrival\} \\
&= \int_{t=0}^{\infty} [\exp(-\lambda_1 t)](\lambda_1 dt) \int_{\alpha=t}^{\infty} [\exp(-\lambda_2 \alpha)](\lambda_2 d\alpha) \\
&= \frac{\lambda_1}{\lambda_1 + \lambda_2}
\end{aligned} \tag{4.74}
$$

$$
\begin{aligned}
P\{I_i < \tau\} &= P\{Class\ 2\ busy\ period\ finishes\ before\ a\ Class\ 1\ arrival\} \\
&= \int_{t=0}^{\infty} f_{BP2}(t) dt \int_{\alpha=t}^{\infty} [\exp(-\lambda_1 \alpha)](\lambda_1 d\alpha) \\
&= \int_{t=0}^{\infty} [\exp(-\lambda_1 t)] f_{BP2}(t) dt \\
&= L_{BP2}(\lambda_1)
\end{aligned} \tag{4.75}
$$

Using Eqs. (4.72)-(4.75) in Eq. (4.71), we get

$$
\begin{aligned}
P\{W_{q1} = 0\} &= \frac{\lambda_1}{\lambda_1 + \lambda_2} \sum_{n=0}^{\infty} \left[\frac{\lambda_2 L_{BP2}(\lambda_1)}{\lambda_1 + \lambda_2} \right]^n \\
&= \frac{\lambda_1}{\lambda_1 + \lambda_2 - \lambda_2 L_{BP2}(\lambda_1)}
\end{aligned} \tag{4.76}
$$

We now consider the case for $WQ1>0$. Recall that this is the case when the class 1 arrival (after the class 1 gets empty) occurs in a I_i interval. Then for $t>0$, the probability density function $f_{WQ1}(t)$ would be given by

$$f_{WQ1}(t)dt = P\{t < W_{q1} \le t + dt\}$$
$$= \sum_{n=0}^{\infty} P\{t < \sum_{i=0}^{n+1}(A_i + I_i) - \tau \le t + dt\} \tag{4.77}$$

Note that in the above, $\left(\sum_{i=0}^{n+1}(A_i + I_i) - \tau\right)$ is the waiting time for this case (i.e. W_{q1}>0)]. Since τ is memory-less, as mentioned earlier, we have

$$P\{t < \sum_{i=0}^{n+1}(A_i + I_i) - \tau \le t + dt\}$$
$$= \left[\prod_{i=0}^{n} P\{A_i < \tau\}P\{I_i < \tau\}\right] P\{A_{n+1} < \tau\}P\{t + \tau < I_{n+1} \le t + \tau + dt\} \tag{4.78}$$
$$= \left[\left(\frac{\lambda_2}{\lambda_1 + \lambda_2}\right)^{n+1} (L_{BP2}(\lambda_1))^n \int_{r=0}^{\infty}[\exp(-\lambda_1 r)] f_{BP2}(r + t)\lambda_1 dr\right] dt$$

Considering both these cases, we get the required probability density function $f_{WQ1}(t)$ for all t as

$$f_{WQ1}(t) = \frac{\lambda_1}{\lambda_1 + \lambda_2 - \lambda_2 L_{BP2}(\lambda_1)}\delta(t)$$
$$+ \frac{\lambda_2}{\lambda_1 + \lambda_2 - \lambda_2 L_{BP2}(\lambda_1)} \int_{r=t}^{\infty} \lambda_1 [\exp(-\lambda_1 (r - t)] f_{BP2}(r)dr \tag{4.79}$$

Taking the transform of both sides of Eq. (4.79) gives the required transform

$$L_{WQ1}(s) = \frac{\lambda_1}{\lambda_1 + \lambda_2 - \lambda_2 L_{BP2}(\lambda_1)} + \frac{\lambda_1 \lambda_2 [L_{BP2}(s) - L_{BP2}(\lambda_1)]}{[\lambda_1 + \lambda_2 - \lambda_2 L_{BP2}(\lambda_1)](\lambda_1 - s)} \tag{4.80}$$

and the mean as

$$\overline{WQ1} = -L'_{WQ1}(s)\Big|_{s=0} = \frac{1 + \lambda_2 (\overline{BP2})}{\lambda_1 + \lambda_2 - \lambda_2 L_{BP2}(\lambda_1)} - \frac{1}{\lambda_1} \tag{4.81}$$

Recall that $\overline{BP2}$ is the mean busy period length for the class 2 queue. Substituting $L_{WQ1}(s)$ from Eq. (4.80) and $L_T(s)$ from Eq. (4.64) in Eq. (4.69), we can obtain $L_{T*}(s)$. Similar substitution in Eq. (4.70) will yield the other generating function $\hat{A}(z)$.

Collecting all the above results for convenience, we can summarise as follows -

*1.*For class 2, the results are the same as that of an M/G/1 queue with average arrival rate λ_2 and service time probability density function $b_2(t)$ with L.T. $L_{B2}(s)$.

2. *For class 1*, solve for $L_{BP2}(s)$ using Eq. (4.57). Use this in Eq. (4.64) to get $L_T(s)$ and in Eq. (4.80) to get $L_{WQ1}(s)$. The generating functions, $A(z)$ and $\hat{A}(z)$ may now be found by substitution in Eqs. (4.67) and (4.70). To find p_0 using Eq. (4.61), the required moments $\bar{a}$ and $\bar{\hat{a}}$ may be obtained from $A(z)$ and $\hat{A}(z)$, respectively, using the moment generating property of the generating functions. These means are obtained to be

$$\bar{a} = \lambda_1 \bar{T} = \lambda_1 \overline{X_1}\left(1 + \lambda_2 \overline{BP2}\right) \tag{4.82}$$

$$\bar{\hat{a}} = \lambda_1\left(\overline{WQ1} + \bar{T}\right) = \bar{a} + \lambda_1 \overline{WQ1} \tag{4.83}$$

After p_0 is found, then using $A(z), \hat{A}(z)$ and Eq. (4.60), the generating function $P_1(z)$ of the number of class 1 users in the system can be found This may then be used to find other performance parameters of the queue for the lower priority class 1 traffic.

4.6 The Discrete Time Geo/G/1 and Geo$^{[X]}$/G/1 Queues

In all our discussions until now and in subsequent chapters of this text, we consider systems, which are continuous in time, where arrivals and departures can occur at any time. In this section we look at one example of a discrete-time system where the time axis is segmented into a sequence of equal time intervals which we will refer to as *slots*. It is assumed that arrivals can only occur at the slot boundaries. Service to a job/customer can also start only at these boundaries and the service duration is always an integral multiple of the slot duration. Consider the arrival process to such a system. We assume that the number of jobs that arrive in successive slots are independent, identically distributed (i.i.d.) random variables. If we assume that only one job can arrive in a slot with probability λ and that no jobs arrive in a slot with probability $1-\lambda$, $(0<\lambda<1)$, then the inter-arrival time I will be geometrically distributed with mean $1/\lambda$ and with probability $P\{I=k$ slots$\}=\lambda(1-\lambda)^{-k}$ for $k=1,2,\ldots\ldots$ This kind of arrival process will be referred to as a *geometric process* or a *Bernoulli process*. A single server queue with infinite buffers and with this arrival process is typically represented as a Geo/G/1 queue. Note that time is measured as multiples of slot durations.

If more than one job can arrive in a slot, then the inter-arrival time between batches will still be considered to be geometrically distributed and the arrivals may be considered to come in batches of random size. A single server queue with infinite buffers of this type will be represented as a $\text{Geo}^{[X]}/\text{G}/1$ queue. Both the Geo/G/1 and the $\text{Geo}^{[X]}/\text{G}/1$ queueing models are useful to analyse service scenarios which operate in a discrete-time mode and are considered in detail in this section.

Consider the $\text{Geo}^{[X]}/\text{G}/1$ queue with Λ as the number of jobs that arrive in a single slot. The probability distribution of Λ will then be specified as

$$\lambda(k)=P\{\Lambda=k\} \qquad k=0,1,2\ldots\ldots\ldots \tag{4.84}$$

with the generating function $\Lambda(z)$ defined as

$$\Lambda(z)=\sum_{k=0}^{\infty}\lambda(k)z^k \qquad |z|\le 1 \tag{4.85}$$

in the usual fashion. Let λ and $\lambda^{(i)}$ respectively denote the mean of Λ and its ith factorial moment as

$$\begin{aligned}\lambda &= E\{\Lambda\}=\Lambda^{(1)}(1)\\ \lambda^{(2)} &= E\{\Lambda(\Lambda-1)\}\\ \lambda^{(i)} &= E\{\Lambda(\Lambda-1)\ldots\ldots(\Lambda-i+1)\}=\Lambda^{(i)}(1) \qquad i=3,4,\ldots\ldots\end{aligned} \tag{4.86}$$

For the special case of the Geo/G/1 queue which allows at the most one arrival per slot, we will have

$$\Lambda(z) = 1 - \lambda(1 - z) \qquad \text{Geo/G/1 arrivals} \tag{4.87}$$

In the discrete-time system, we need to be more precise about the exact arrival points of jobs arriving in a slot and the slot in which service to a job, which arrives when the system is empty, actually gets started. This leads to the two choices - *late arrival model* and *early arrival model* - described subsequently. Note that we will always measure the queue size immediately after the slot boundaries and that this will not be affected by the choice of the early/late models. The measurement of the waiting time will, however, depend on the choice of this model.

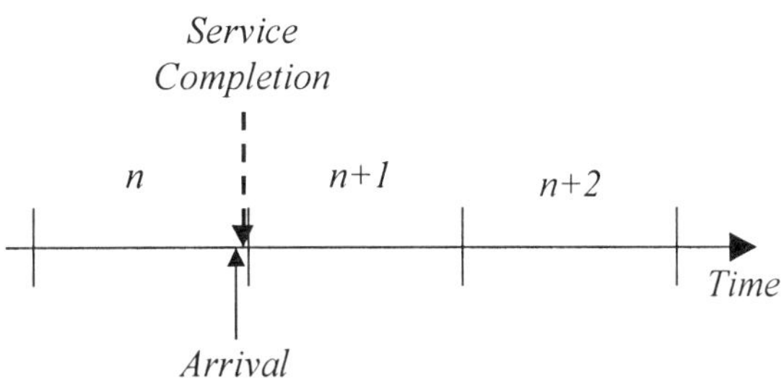

Figure 4.6. Late Arrival Model of a Discrete Time Queue

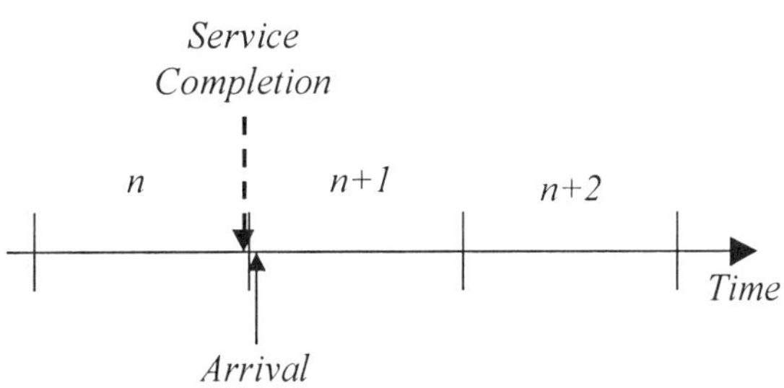

Figure 4.7. Early Arrival Model of a Discrete Time Queue

In the *late arrival model* shown in Figure 4.6, jobs are assumed to arrive late in the slot, i.e. just before the slot ends. The arrivals, if any, will then see the job, which has just finished service about to leave, and this departure will include the newly arrived jobs in the ones it leaves behind in the queue. In this case, if the new job arrives to an empty queue (say at the end of the nth slot), then it will enter service from the beginning of the next (i.e. the $(n+1)^{th}$ slot). In this case, the time spent in the queue is measured from the $(n+1)^{th}$ slot, i.e. the slot after the one in which the job actually arrives.

In the *early arrival model* shown in Figure 4.7, we assume that the jobs arrive early in the slot, i.e. just after the slot begins. If the queue is empty when the arrival occurs, service to the job starts immediately from the same slot. This slot (where the job arrives early) is also counted in measuring the time spent in the queue by the arriving job.

A close examination of the early and late arrival models lead to some useful conclusions. For a queue in steady state, the joint distributions of the queue size and the residual or elapsed service time at an arbitrary point on the continuous time axis will be the same in both the models. The waiting time in queue for a particular job will also be the same in both the models. The queue size at service completion in an early arrival model will, however, be lower than that of the corresponding late arrival model by the number of jobs that actually arrive at that slot boundary. Note that the queue size is always measured immediately after a slot boundary (as mentioned earlier) which gives rise to the above difference between the two models.

In our analysis here, we assume that the service duration can have any general distribution as long as the service time of a job is an integral multiple of the slot duration. Service starts and ends only at the slot boundaries. Let X be the (random) service time of a job measured in units of slots, i.e. the service duration is of duration X slots. The probability distribution $b(k)$ of X is given by

$$b(k) = P\{X = k\} \qquad k=1,2.......... \tag{4.88}$$

with the generating function $B(z)$ defined as

$$B(z) = \sum_{k=1}^{\infty} b(k) z^k \qquad |z| \leq 1 \tag{4.89}$$

Let b and $b^{(i)}$ respectively denote the mean of X and its ith moment as

$$\begin{aligned} b &= E\{X\} \\ b^{(2)} &= E\{X^2\} \\ b^{(i)} &= E\{X^i\} \qquad i = 3,4,........ \end{aligned} \tag{4.90}$$

The load offered to the single server queue is defined in the usual manner to be given by ρ, with $\rho = \lambda b$. For a stable queue, we require that the offered load be less than unity, i.e. $\rho < 1$. This stability condition will be assumed in our subsequent analysis.

4.6.1 The Geo/G/1 Queue

We consider here the case where only one arrival, at the most, can occur in a slot and the service duration of a job is an integral (random) multiple of the slot duration. The queue has only one server and has an infinite number of waiting positions. We consider the analysis of this queue separately for the late arrival and early arrival models.

Late Arrival Model

Let n_i be the number of jobs in the queue immediately after the service completion of the ith job. Let a_i be the number of jobs arriving during the service time of the ith job. This leads to the following for the late arrival model

$$\begin{aligned} n_{i+1} &= a_{i+1} & n_i &= 0 \\ &= n_i + a_{i+1} - 1 & n_i &\geq 1 \end{aligned} \tag{4.91}$$

The random variables a_i $i=1,2,......$ are independent and identically distributed random variables with the generating function $A(z)$ and mean ρ. The state of the queue n_i will then form a homogenous Markov chain. Under conditions of equilibrium, the steady-state distribution p_k of this Markov chain will be

$$p_k = \lim_{i \to \infty} P\{n_i = k\} \qquad k = 0,1,2,....... \tag{4.92}$$

with the generating function $P(z)$ found using

$$P(z) = \sum_{k=0}^{\infty} p_k z^k \tag{4.93}$$

Note that Eq. (4.91) is in the same form as Eq. (3.11), which was the equivalent state expression derived for the M/G/1 queue. It can therefore be solved in the same fashion as Section 3.2 to find $P(z)$ as

$$P(z) = \frac{(1-\rho)(1-z)A(z)}{A(z)-z} \tag{4.94}$$

with

$$p_0 = 1 - \rho \tag{4.95}$$

As in Section 3.2, *A(z)* is the generating function for the number of jobs arriving during a service interval. Using Eq. (4.87), it can be shown that for the Geo/G/1 queue we will get

$$A(z) = \sum_{j=1}^{\infty} b(j) \sum_{k=0}^{j} \binom{j}{k} \lambda^k (1-\lambda)^{j-k} z^k = B(1-\lambda+\lambda z) \tag{4.96}$$

This leads to

$$P(z) = \frac{(1-\rho)(1-z)B(1-\lambda+\lambda z)}{B(1-\lambda+\lambda z)-z} \tag{4.97}$$

as the generating function for the number in the system at the instant of service completion. In this case also, it can be shown as in Section 3.2, that the generating function of the queue states will be given by *P(z)* even at the time instant immediately after each slot boundary. (This will actually also be true at any arbitrary time instant on the continuous time axis.). Therefore, we can use *P(z)*, as given in Eq. (4.97), to find the equilibrium state distribution of the queue (i.e. the probability distribution of the number in the system) immediately after each slot boundary.

The discrete-time equivalent of the PASTA property may also be proved. This property is sometimes referred to as BASTA (i.e. Bernoulli arrivals see time averages) or GASTA (geometric arrivals see time averages). Using this property, we can then also claim that the state distribution as given by *P(z)* of Eq. (4.97) will also hold for the number in the system as seen by an arriving customer.

Using *P(z)* from Eq. (4.97), the mean number N in the system may be obtained to be

$$N = \frac{\lambda^2 b^{(2)} - \lambda\rho}{2(1-\rho)} + \rho \tag{4.98}$$

The discrete-time version of Little's result also holds and is stated as

$$N = \lambda W \tag{4.99}$$

where N is the mean number in the system as given by Eq. (4.98), λ is the mean number of arrivals in one slot and W is the mean time spent in system (in units of slots) by an arriving customer. This may then be used to obtain the mean time spent in system. Once W has been obtained, the mean time W_q spent waiting for service (in number of slots) will be given as

$$W_q = W - b = \frac{\lambda b^{(2)} - \rho}{2(1-\rho)} \tag{4.100}$$

Little's result ($N_q = \lambda W_q$) may then be applied once again to find the mean number waiting in queue in the system, prior to service.

Other results, similar to the ones derived in Chapter 3 for the M/G/1 queue may also be similarly derived for this Geo/G/1 queue. For example, consider a FCFS Geo/G/1 queue. In this case, the number left behind in the system by a departing customer will be the same as the number arriving to the system, while that customer was in service. Let $G_W(z)$ be the generating function for the number of slots for which an arriving job stays in the system. We can then show that

$$P(z) = G_W(1 - \lambda + \lambda z) \tag{4.101}$$

This will allow us to find the required generating function $G_W(z)$ as

$$G_W(z) = \frac{(1-\rho)(1-z)B(z)}{(1-z) - \lambda(1 - B(z))} \tag{4.102}$$

Since the waiting time in queue for a job will be independent of its service time (each one of them measured in units of slots), we can write

$$G_W(z) = G_{W_q}(z)B(z) \tag{4.103}$$

where $G_{Wq}(z)$ is the generating function of the number of slots spent waiting in queue by a job before its service actually starts. Therefore

$$G_{W_q}(z) = \frac{(1-\rho)(1-z)}{(1-z) - \lambda(1 - B(z))} \tag{4.104}$$

Early Arrival Model

In this case, let the state n_i be the system state after the completion of the ith service and before the possible arrival point. Let the number of arrivals a_i in a service duration be defined as for the late arrival case. At equilibrium, the generating functions for these are *P(z)* and *A(z)*, respectively. In addition we define $\widetilde{a}_i$ as the "number of jobs arriving in the service time of the ith job minus one slot" with equilibrium values of its probability distribution and generating function given by

$$P\{\widetilde{a}=k\}=\sum_{j=k+1}^{\infty}\binom{j-1}{k}\lambda^k(1-\lambda)^{j-1-k}b(j) \qquad k=0,1,2,.......$$

$$\begin{aligned}\widetilde{A}(z)&=\sum_{k=0}^{\infty}z^k P\{\widetilde{a}=k\}\\&=\sum_{k=0}^{\infty}z^k\sum_{j=k+1}^{\infty}\binom{j-1}{k}\lambda^k(1-\lambda)^{j-1-k}b(j)\\&=\sum_{j=1}^{\infty}b(j)\sum_{k=0}^{j-1}\binom{j-1}{k}(\lambda z)^k(1-\lambda)^{j-1-k}\\&=\sum_{j=1}^{\infty}b(j)(1-\lambda+\lambda z)^{j-1}\\&=\frac{B(1-\lambda+\lambda z)}{1-\lambda+\lambda z}\end{aligned} \tag{4.105}$$

The state transition equation (corresponding to Eq. (4.91) for the late arrival case) may then be written as

$$\begin{aligned}n_{i+1}&=\widetilde{a}_{i+1} && n_i=0\\&=n_i+a_{i+1}-1 && n_i\geq 1\end{aligned} \tag{4.106}$$

This leads to the following generating function for the system states at equilibrium

$$P(z)=\frac{(1-\rho)[A(z)-z\widetilde{A}(z)]}{(1-\lambda)[A(z)-z]} \tag{4.107}$$

Simplifying Eq. (4.107) by substitution using Eqs. (4.96) and (4.105), we get the final expression for the generating function $P(z)$ as

$$P(z) = \frac{(1-\rho)(1-z)B(1-\lambda+\lambda z)}{(1-\lambda+\lambda z)[B(1-\lambda+\lambda z)-z]} \tag{4.108}$$

Comparing this with the result of Eq. (4.97) for the late arrival model, we see that the expressions differ merely by a scaling factor of $(1-\lambda+\lambda z)$. The queue size in the early arrival model is lower than in the late arrival model. This happens because in the early arrival model, we are observing the queue size before the possible arrival point in a slot. Note that since the slot in which the job arrives is counted in the early arrival model, when we calculate the time spent by it in the queue, the expression for $G_W(z)$ given in Eq. (4.101) for the late arrival model will get modified to be

$$P(z) = \frac{1}{(1-\lambda+\lambda z)} G_W(1-\lambda+\lambda z) \tag{4.109}$$

Using this, the generating function $G_W(z)$ for the number of slots spent in the system by a job (waiting and in service) may be obtained as

$$G_W(z) = \frac{(1-\rho)(1-z)B(z)}{(1-z)-\lambda(1-B(z))} \tag{4.110}$$

Note that this is the same as the result of Eq. (4.102) for the late arrival model. This is expected since the number of slots that are spent in the system by a job will be the same in both cases. Other results, such as those obtained for the late arrival model, may also be similarly obtained.

4.6.2 The Geo$^{[X]}$/G/1 Queue

We consider here the case where there may be more than one job arrival in a slot. In all other aspects, the queue is the same as that examined for the Geo/G/1 case. The generating function for the number of arrivals in a slot is given by $\Lambda(z)$ as in Eq. (4.85) with $\lambda(k)$ as the probability of k arrivals in a slot. As for the Geo/G/1 case, for both the late and early arrival models, we consider the distribution of the queue size (i.e. its generating function $P(z)$) immediately after a service completion. For the distribution of the waiting time or of the time spent in the system, we consider this immediately after each slot boundary for the late arrival model. In contrast, for the early arrival model, we consider this immediately after a possible arrival following the slot boundary.

Late Arrival Model

Let n_i be the number of jobs in the queue immediately after the service completion of the ith job. Let a_i be the number of jobs arriving during the service time of the ith job. We also assume that Λ^* is the number of jobs arriving in a slot which is such that there is at least one job arrival in that slot. Using the notation of Eqs. (4.84) and (4.85), the generating function $\Lambda^*(z)$ for this may be derived to be

$$\Lambda^*(z) = \frac{\Lambda(z) - \lambda(0)}{1 - \lambda(0)} \tag{4.111}$$

The state transition equations corresponding to the Markov chain of n_i, $i=1,2,\ldots\ldots$ may be written as

$$\begin{aligned} n_{i+1} &= \Lambda^* + a_{i+1} - 1 \qquad & n_i = 0 \\ &= n_i + a_{i+1} - 1 \qquad & n_i \geq 1 \end{aligned} \tag{4.112}$$

Using the fact that, under equilibrium conditions, the generating function $A(z)$ of the number a arriving within a service duration will be given in this case by

$$A(z) = B(\Lambda(z)) \tag{4.113}$$

we get the generating function $P(z)$ of the number in the system immediately after the service completion instants as

$$P(z) = \frac{(1-\rho)[1-\Lambda(z)]B(\Lambda(z))}{\lambda[B(\Lambda(z)) - z]} \tag{4.114}$$

In this case, the generating function for the number in the system immediately after an arbitrary slot boundary is not the same as $P(z)$. (Note that the two were equal for the Geo/G/1 queue.) [BrK93] claims that, at equilibrium, the following relationship between the generating function at service completion and the generating function $P^*(z)$ immediately after arbitrary slot boundaries will hold for a large class of discrete time models, including the $\mathrm{Geo}^{[X]}$/G/1 queue. This relationship is expressed as

$$P(z) = P^*(z)\frac{1-\Lambda(z)}{\lambda(1-z)} \tag{4.115}$$

Note that substituting $\Lambda(z)=1-\lambda+\lambda z$ in Eq. (4.115) gives $P(z)=P^*(z)$ for the Geo/G/1 queue as per our earlier claim that for this case, the two generating functions will be equal.

Using Eq. (4.115), we get that the generating function of the system state immediately after an arbitrary slot boundary will be given by

$$P^*(z) = \frac{(1-\rho)(1-z)B(\Lambda(z))}{B(\Lambda(z))-z} \tag{4.116}$$

Since nothing in the system can change between the slot boundaries, the system state at any arbitrary point in time will then also be equal to the state observed (i.e. the generating function giving the state probabilities) at the preceding slot boundary. We can also use the generating function $P^*(z)$ of Eq. (4.116) to directly compute the mean number N in the system just after any arbitrary slot boundary (or at any arbitrary time instant) to be

$$N = \frac{\lambda^2 b^{(2)} - \lambda\rho + \lambda^{(2)} b}{2(1-\rho)} + \rho \tag{4.116}$$

Even if we assume a FCFS service discipline as before, deriving the delay distribution for this case requires more effort than the corresponding case of the Geo/G/1 queue. This is because one would need to account for the fact that arrivals come in batches, which may have more than one job in them. Consider a particular job in a batch whose queueing delay (prior to service) is required to be found. It is convenient to tackle this by considering the total queueing delay w_q of a job in a batch (mean W_q with generating function $G_{Wq}(z)$) to consist of the batch queueing delay w_b (mean W_b with generating function $G_{Wb}(z)$) and a delay w_x (mean W_x with generating function $G_{Wx}(z)$) corresponding to the service time of those members of the batch who get served before the job of interest. (The batch queueing delay is the time to wait before service to the first member of the batch can begin.). Since the two components of the queueing delay will be independent, we observe that

$$\begin{aligned} W_q &= W_b + W_x \\ G_{Wq}(z) &= G_{Wb}(z)G_{Wx}(z) \end{aligned} \tag{4.117}$$

Note that this argument is essentially similar to the approach taken to handle the queueing delay in the $M^{[X]}/G/1$ queue in Section 4.4. Following an approach similar to the one used there, we can derive the generating functions $G_{Wb}(z)$ and $G_{Wx}(z)$ and the corresponding means as given next.

$$G_{Wb}(z) = \frac{(1-\rho)(1-z)}{\Lambda(B(z))-z} \qquad W_b = \frac{\lambda b^{(2)} - \rho + \lambda^{(2)} b^2}{2(1-\rho)} \tag{4.118}$$

$$G_{Wx}(z) = \frac{1-\Lambda(B(z))}{\lambda[1-B(z)]} \qquad W_x = \frac{\lambda^{(2)} b}{2\lambda} \tag{4.119}$$

This leads to

$$\begin{aligned} G_{Wq}(z) &= \frac{(1-\rho)(1-z)[1-\Lambda(B(z))]}{\lambda[\Lambda(B(z))-z][1-B(z)]} \\ W_q &= \frac{\lambda^2 b^{(2)} - \lambda\rho + \lambda^{(2)} b}{2(1-\rho)} \end{aligned} \tag{4.120}$$

as the generating function and mean of the queueing delay encountered by a job arriving to the system (as possibly part of a batch). The relationship between the mean queueing delay W_q and the mean time W spent in the system by an entering job will still be given as $W=W_q+b$ where b is the service time of a job, as defined earlier. This leads to

$$W = \frac{\lambda^2 b^{(2)} - \lambda\rho + \lambda^{(2)} b}{2(1-\rho)} + b \tag{4.120}$$

which agrees with value one would obtain by applying Little's result to Eq. (4.116). Applying Little's result to (4.120), we can also get the mean number of jobs N_q waiting in queue (prior to service).

Early Arrival Model

In this case, we consider the system state n_i immediately after the ith service completion and before a possible arrival at the beginning of the following slot. These will form a Markov chain with the transitions governed by the following

$$
\begin{aligned}
n_{i+1} &= \Lambda^* + a'_{i+1} - 1 \qquad & n_i = 0 \\
&= n_i + a_{i+1} - 1 \qquad & n_i \geq 1
\end{aligned}
\tag{4.121}
$$

Here Λ^* is defined as for the case of the late arrival Geo/G/1 queue with generating function $\Lambda^*(z)$ given by Eq. (4.111). The number of arrivals in the $(i+1)^{th}$ service duration is a_{i+1} with its generating function $A(z)$ given by Eq. (4.113) under equilibrium conditions. The quantity a'_{i+1} represents the number of jobs arriving during the $(i+1)^{th}$ service time given that the system state $n_i = 0$, i.e. the ith departure left the system empty. This will also be the number of arrivals during the last $k-1$ slots if the service time of the $(i+1)^{th}$ job is k slots. Under equilibrium conditions, the generating function $A'(z)$ for this may be derived as

$$
A'(z) = \sum_{k=1}^{\infty} b(k)[\Lambda(z)]^{k-1} = \frac{B(\Lambda(z))}{\Lambda(z)} \tag{4.122}
$$

Using these the generating function $P(z)$ of the system state at the designated time instants will be

$$
P(z) = \frac{(1-\rho)[1-\Lambda(z)]B(\Lambda(z))}{\lambda\Lambda(z)[B(\Lambda(z)) - z]} \tag{4.123}
$$

In this case, as well, the generating function $P^*(z)$ of the number in the system immediately after an arbitrary slot boundary (or at any arbitrary point on the continuous time axis) will be different from the generating function $P(z)$ given above in Eq. (4.123). As a matter of fact, it can be shown that the expression for $P^*(z)$ will be the same as that given for the late arrival model in Eq. (4.116). Since the values of the queue's output parameters N, N_q, W and W_q are derived using the generating function $P^*(z)$ and Little's result, the expressions for these parameters for the early arrival model will also be the same as for the late arrival model.

We have considered only two basic discrete-time queues in this section. More detailed treatment of such queues may be found in [Tak94] and [BrK93].

Problems

1. Consider the M/G/1 queue with exceptional first service discussed in Section 4.3. Analyse this queue using the residual life approach assuming the following -

> The basic service time X has mean $\overline{X}$ and second moment $\overline{X^2}$. The additional service time required by the first customer in the busy period is Δ with mean $\overline{\Delta}$ and second moment $\overline{\Delta^2}$. The random variables X and Δ are independent and identically distributed random variables.

2. Analyse a M/G/1 queue such that whenever there are N or more customers in the system at the beginning of a service time, the service time distribution is modified so that its L.T. is $B^*(s)$ rather than the normal $B(s)$. Set up the approach that would be needed to solve for a general value of N but solve explicitly only for $N=2$.

3. Analyse a M/G/1 queue where the arrivals come from a Poisson process with rate λ_I when the server is idle and with rate λ_B from a Poisson process when the server is busy.

4. Consider a $M^{[X]}/G/1$ queue with a service facility whose service time distribution is given in terms of the L.T. of its pdf as

$$L_B(s) = \frac{\mu(s+\mu)}{s^2 + 3\mu s + \mu^2}$$

The batch arrival process generates batches at rate λ where the batches are either of size *1* or *2* with equally likely probabilities. Solve this system using the standard approach of Section 4.4 to obtain the corresponding results.

5. For the $M^{[X]}/G/1$ queue of Problem 4, we consider the service facility in more detail. Assume that the service facility contains two stages of exponential servers, where stage 1 serves at rate 2μ and stage 2 serves at rate μ. A job entering the service facility first gets served at stage 1. On completing this service, the job leaves the system with probability *0.5* or joins service at stage 2 with probability 0.5. On completing service at stage 2, the job enters service at stage 1 once again and goes through the same random choice as described above once it finishes service at stage 1. This continues until the job leaves the system.

Show that this model of the service facility will lead to the same overall service time distribution as given in Problem 4. Note that with the service modelled in terms of the stages as described above, we can now also use the method of stages to solve this queue. Set up the state transition diagram and the balance equations that will be required to solve the queue in this fashion.

The approach given by the method of stages may also be used if the queue has a finite capacity. Use this approach to solve the $M^{[X]}/G/1/2$ queue of this type when either a whole batch acceptance strategy or a partial batch acceptance strategy is used. For both these strategies, calculate the probability that a job arriving to the system is actually allowed to enter the queue.

6. Derive results similar to the ones derived in Section 4.4 for the $M^{[X]}/G/1$ queue where the batch sizes are geometrically distributed. Assume that the probability that a batch will be of size n will be given by $(1-q)q^n$ for $n \geq 0$. Use standard assumptions and notations otherwise.

7. Students enter the dining hall for breakfast in equally likely groups of either one or two with a group arrival rate of λ. The first member of the group is served in an exponentially distributed time X with probability density function $b(t)$ and L.T. $\tilde{B}(s)$. The second member (if any) orders an extra omelette which requires Δ seconds more where Δ is fixed. The mess operates as a single server $M^{[X]}/G/1$ queue. Find the mean delay that an arriving student will encounter before being served.

8. Consider an $M^{[X]}/G/1$ queue where the first customer in the batch requires exceptional service which is Δ seconds more than the normal service time. Obtain the queueing delay for a customer in this system. Use standard assumptions and notations, otherwise.

9. Consider an $M^{[X]}/G/1$ queue with vacations and carry out the corresponding analysis. This can be tried with either a normal vacation model (repetitive vacations on idle) or a single vacation per idle model.

10. Consider a 2-priority preemptive resume priority M/G/1 queue with high priority customers of class 2 and lower priority customers of class 1. The system enforces the rule that there can be only one class 2 customer in the system at any time (i.e. there is no buffering for class 2) - however, there is *infinite buffering* for class 1 customers. Let X_2 (fixed, not random) be the service time for class 2 and X_1 (fixed, not random) be the service time for class 1. Arrival processes are Poisson with average arrival rate λ_1 for class 1 and λ_2 for class 2. For this, consider a class 1 customer who start service at time t and leaves the system at time $t+T$. What would be the distribution (or L.T.) of the random variable T?

11. Consider the system of Problem 10 once again except that we now assume that class 2 customers can also be infinitely buffered. For this obtain the average total delays encountered by class 2 and class 1 customers.

12. For the n-priority Non-preemptive M/G/1 queue, show that

$$\sum_{k=1}^{n} \rho_k W_q^{(k)} = \frac{R\rho}{(1-\rho)}$$

where $W_q^{(k)}$ is the mean wating time in queue for a customer of priority class k. This is an example of the so-called *Work Conservation Principle*!

Chapter 5

Fundamentals of Queueing Networks

Open and Closed Networks with Product-Form Solutions

In this chapter, we consider service models, where the service to be provided may be represented as a sequence of services provided by several servers. Such a service scenario may be conveniently represented as *Networks of Queues* as shown in Figures 5.1 and 5.2. In these models, customers/jobs which finish service at a queue may either move on for the next stage of service to another queue (or even re-enter the earlier queue) or may leave the network altogether.

A standard example of such a queueing network type model arises in modelling a machine shop where there are a number of machines, each with its own queue. A job here would require a sequence of operations by one or more machines in sequence. A job entering the machine shop would correspond to arrivals to this network model and the departures from the shop occur when all the required services at the designated machines have been obtained. The types of operations to be done decide the machines that the job must go through while it is in the machine shop and the sequence in which these operations are needed to be done at the various machines involved decide the routing of jobs from one machine to another.

A communication/computer network with various service facilities and store and forward nodes may also be modelled similarly as a network of queues. The individual packets or messages that have to be carried from their respective source nodes to their corresponding destination nodes correspond to the jobs or customers that have to be served in the system. These enter the system from their respective source nodes and leave the system from their designated destination node. During their transit through the system, they may have to go through store and forward queues at various intermediate nodes leading naturally to a queueing network based model for the required analysis.

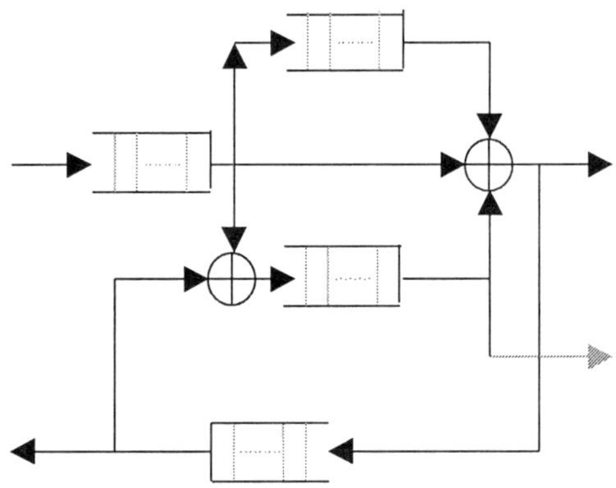

Figure 5.1. An Open Queueing Network

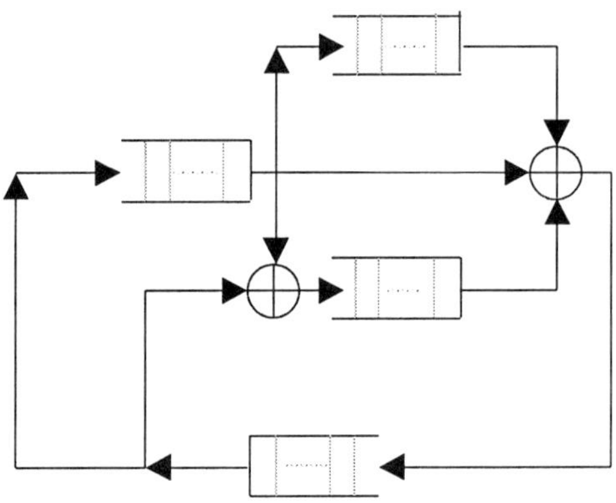

Figure 5.2. A Closed Queueing Network

Modelling a service scenario as a queueing network, rather than as a single queue, is also a very convenient way of representing it and may lead to simple analytical techniques that may be conveniently applied. The *Method of Stages* described in Section 2.9 for analysing $M/E_K/1$ is an example of such a technique where the Erlangian service time is decomposed into two or more successive exponential service times to make the analysis simpler. This model did require some special restrictions on the movement of jobs as new jobs were not allowed to enter the first stage unless the previous job has exited from the system after completing its service at all the intermediate stages. The queueing networks considered by us here (and in Chapter 6) are not as restrictive as this and indeed allow a great degree of flexibility in routing jobs from one queue to another. As a result, some very complex queueing scenarios can be modelled in this fashion.

In order to keep our models as general as possible, we also assume that routing of jobs from the output of a queue can be probabilistic in nature. This implies that a job finishing its service at one queue, may get routed to more than one queue with the actual queue to which it will be routed chosen probabilistically using a pre-defined probability parameter, i.e. it will be routed to the queue Q_k with probability p_k for the various queues $Q_1,......,Q_K$ in the network. This is referred to as *Probabilistic Routing* and will be the routing model used by us in this chapter and the next. It is also possible to have *Deterministic Routing* where the jobs are deterministically routed from one queue to another. In the case of multi-class jobs, jobs of different classes may have different deterministic routes that they follow through the queueing network between the queue where they enter and the queue from where they leave the network. In the case of deterministic routing, a job of a particular class, entering the network at some node, may be required to follow a definite sequence of nodes before its eventual departure from the network. On the other hand, if the routing is probabilistic then a job may have to probabilistically (i.e. randomly) decide which node to go to (out of a set of nodes), after it finishes service at the previous node/queue. In the case of probabilistic routing, the routing choices made by individual jobs are also assumed to be independent in nature. This implies that the choice of a particular path by a job does not affect the way other jobs (following probabilistic routing) will choose their paths through the network.

The techniques described in this chapter provide the basic tools to analyse simple queueing networks provided some important assumptions about the nature of the network are satisfied. Subsequently, in Chapter 6, we consider approximate methods that give good results even when some of these assumptions are not strictly met by the queueing network. For more detailed advanced treatment of queueing networks, we refer the readers to [Per94], [Wal88] and [Woo89] and various other research papers which will be referenced subsequently at appropriate places in Chapters 5 and 6.

5.1 Classification of Different Types of Queueing Networks

The most common method of classifying queueing networks is to consider them as being either *open* or *closed* networks depending on whether jobs are allowed to enter and leave the system or whether the system always has a fixed number of jobs circulating in it. It is also possible to define a *mixed* queueing network when one is considering a situation where there are a number of different classes of jobs in the system. In this case, the network may be open for some of the classes while it may be closed for some other classes of customers.

5.1.1 Open Queueing Networks

An *Open Network* is one where jobs/customers arrive from outside to one or more queues and eventually leave the network from some of the queues. Inside the network, a job finishing service at one queue may leave the network altogether or may go to another queue for service. We will consider this routing to be probabilistically done in our subsequent discussions but the routing can actually also be deterministic in nature for multiple job classes. Arrivals from the external world to the network may take place in one or more than one queues and departures may leave the network from one queue or more than one queue. An example of an open network has been shown in Figure 5.1.

In addition, a open network is considered to be *Acyclic* if a node is visited by a job at most once during its entire sojourn through the network. Since there is no feedback in such an acyclic network, it is also sometimes referred to as a *Feedforward Network.* This implies that there will be no feedback in such an open network. If there is feedback present in an open queueing network, then it may lead to situations where a job finishing service at one queue may return to the same queue later for another round of service.

5.1.2 Closed Queueing Networks

A *Closed Network* is one where there are a constant number of jobs that continually circulate in the network with no other arrivals to the system or departures from the system. The number of jobs in the network is fixed and the sum of the jobs at all the individual queues always equals the total number of jobs in the system. No additional jobs can enter the network through any of its component queues. On completion of service at a particular queue, a job will get routed to another queue in the network. This is done just as in the case of open networks, except that jobs cannot be routed out of the network but will have to be routed to another queue, either probabilistically or in a deterministic fashion.

A closed network seems an unusual one at first sight since it does not allow jobs to enter or leave the network. In reality, this may actually be a good way of modelling a service facility. A typical system of this type would be one where there is a large (infinite) number of jobs waiting to enter a system at all times, the number of jobs allowed inside the system is fixed at some value and a job enters the system immediately, whenever the sequence of services for another job is completed. This, for example, is a typical way of modelling a time-shared computer system where a fixed number of jobs, individually ask for a number of computing and/or input/output services that they require for their task. Whenever the tasks required for a job are finished, another job enters the system for its required services, replacing the earlier one, which has got the services it required.

5.1.3 Mixed Queueing Networks

As mentioned before, one sometimes also refers to a *Mixed Network* where there are multiple classes of jobs. For some of these job classes, the network will be closed (i.e. no arrivals from the outside or departures to the outside for these classes) while it would be open for the other job classes (i.e. jobs of these classes arrive from outside the network and eventually leave the network).

5.1.4 Multiple Classes of Jobs/Customers

One can have open or closed networks with multiple classes as well. A *Multi-class Open Network* will have a number of classes of customers with their respective routing and arrival/service specifications where the network is open with respect to each of these classes. In a *Multi-class Closed Network*, the network will be closed for each one of these classes even though individual classes would have their own routing and service specifications.

5.1.5 Blocking in a Queueing Network

Queueing networks may be classified in other ways as well. For example, a queueing network where all queues are of infinite capacity is categorized as a *Queueing Network without Blocking*; here, a job can never be blocked while moving from one queue to another (i.e. because the destination queue is full). If a queueing network has one or more queues of finite capacity then it will be classified as a *Queueing Network with Blocking* [Per94]. In such a network, a job, which wants to move to a finite capacity destination queue, will experience blocking if the destination queue is full and cannot accommodate the job trying to enter. Special rules will need to be formulated to handle this blocking event. Depending on these rules, different types of blocking will be classified subsequently in Chapter 6.

5.1.6 Product-Form Queueing Networks

For analyzing a queueing network under steady state conditions with appropriate simplifying conditions, one can model the state of the system as a vector consisting of the states of the individual queues. The system may then be modeled as a *Multidimensional Birth-Death* process and analyzed. In a large number of cases, under fairly general conditions or as a good approximation, the solution for the overall system state may be represented by a *Product Form Solution* of the following form.

$$P\{n_1, n_2, \ldots\ldots\ldots n_K\} = \prod_{i=1}^{K} f_i(n_i)$$

Here, $P\{n_1, n_2, \ldots\ldots, n_K\}$ is the joint distribution of the states in the K queues in the network. The existence of the product form solution implies that this joint distribution may be written as a product of terms which are functions of the individual queue states n_i, i.e. $\{f_i(n_i)\}$, $i=1,\ldots\ldots,K$. (There may be an additional normalisation constant term ensuring that the sum of the probabilities of all the network states sum to unity.) We consider networks of queues where either this property holds exactly or where it may be considered to hold as a good approximation.

5.2 Probabilistic Routing in a Queueing Network

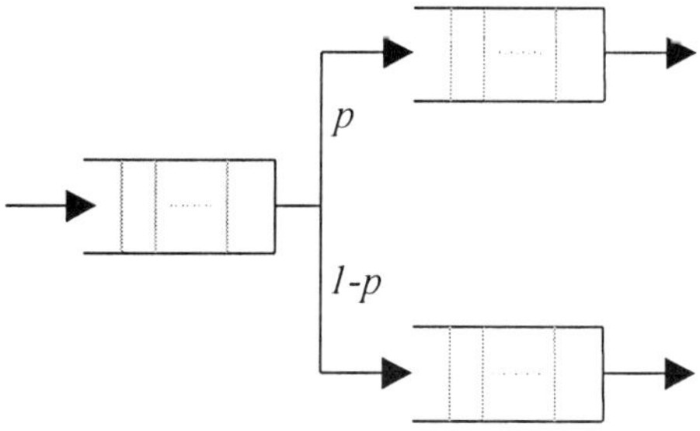

Routing Probabilities are p and $(1-p)$

Figure 5.3. Probabilistic Routing in a Queueing Network

Consider a queueing network with K queues, where a job leaving Q_i can move to any one of queues $Q_1, \ldots\ldots Q_K$ (or can leave the system in the case of an open network). An example of such a network with three queues is shown in Figure 5.3. We would typically consider *Probabilistic Routing* where the job leaving Q_i decides randomly which destination it would go to and this random decision is made independently for each job leaving Q_i. In such a network, a probability matrix will be needed to specify the routing probabilities with which jobs move from one queue to another or leave the system. In this text, we will only consider this kind of probabilistic routing strategy. The i-j^{th} entry, p_{ij}, of the transition probability matrix (also called the *routing matrix*) denotes the probability that a job finishing service at Q_i decides to go to Q_j in the network.

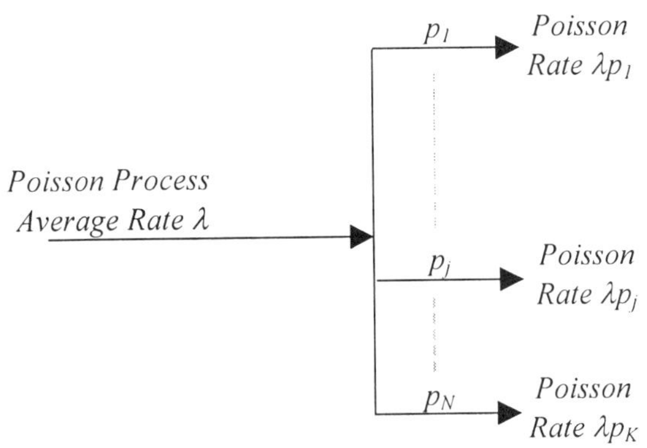

Figure 5.4. Probabilistically Splitting a Poisson Process

Probabilistic routing may also be viewed as randomly splitting a process into several processes. In case the original process is Poisson and if the probabilistic splitting is done in a manner which is independent of the inter-arrival times of the original arrival process, then one can easily show that the processes obtained after the splitting will also be Poisson in nature. As an example consider the case of binary splitting, where on getting an arrival from a Poisson process with average rate λ, we toss a coin which has $P\{head\}=p$ and $P\{tail\}=1-p$. The arrival is sent on route 1 on getting a head and is sent on route 2 on getting a tail. Note that this effectively amounts to splitting the original Poisson process probabilistically with probabilities p and $1-p$ to get two processes with average rates λp and $\lambda(1-p)$. It can be easily shown that if independent random choices with probabilities p and $(1-p)$ are made for each arrival then the two processes obtained after splitting are independent of each other and are also Poisson in nature with the average rates as given earlier. This may be generalized to state that the processes obtained after splitting probabilistically a Poisson process with average rate

λ, using probabilities $\{p_i\}$, would also be independent Poisson processes with rates $\{\lambda p_i\}$. We have illustrated this kind of splitting in Figure 5.4.

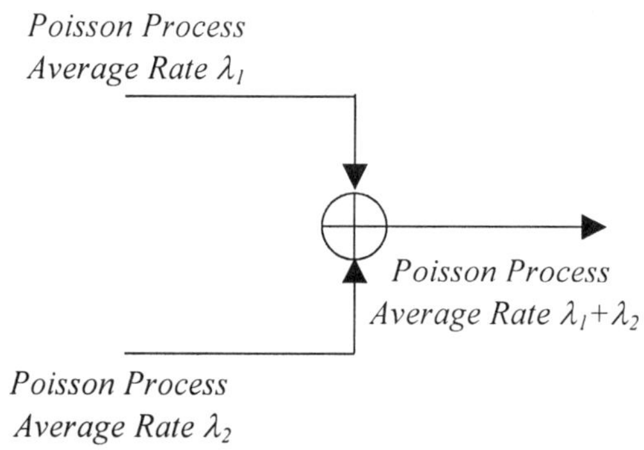

Figure 5.5. Combining Poisson Processes

In this context, it is also useful to note that combining Poisson processes with different rates will also lead to a Poisson process with a rate equal to the sum of the individual rates, i.e. the sum of *N* Poisson processes with average rates λ_1, ..., λ_N is also a Poisson process with average rate (λ_1 + ...+ λ_N). This is illustrated in Figure 5.5 for the case where two Poisson processes of rates λ_1 and λ_2 combine to give a Poisson process of rate $\lambda_1+\lambda_2$.

5.3 Open Networks of M/M/m Type Queues and Jackson's Theorem

It may be recalled that Burke's Theorem (Section 2.7) states that, under equilibrium conditions, the departure process from a M/M/m/∞ queue will also be a Poisson process with the same average rate as the arrival rate to the queue. This may be used in conjunction with the following results,

(a) splitting a Poisson process randomly yields Poisson processes after splitting
(b) sum of Poisson processes is also a Poisson process

to analyse open networks of M/M/m/∞ queues which are acyclic in nature, i.e. feedforward networks without any feedback.

As an example of this, we may consider the queueing network shown in Figure 5.6. Consider the network shown with inputs from Poisson processes with average rates $\lambda 1$ and $\lambda 2$ at Q_1 and Q_2, respectively. The routing probabilities have also been shown.

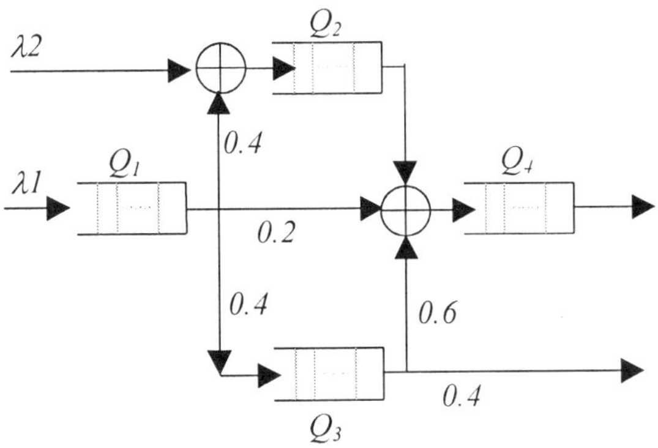

Figure 5.6. A Feedforward Open Network of M/M/m Queues

Using these routing probabilities and the fact that flow balance will exist at equilibrium, we can obtain the average rate of arrivals entering each of the queues. Let λ_{Q1}, λ_{Q2}, λ_{Q3} and λ_{Q4} be the respective average arrival rates of the jobs actually entering queues Q_1, Q_2, Q_3 and Q_4. These will be given by

λ_{Q1} = Average arrival rate for $Q_1 = \lambda 1$
λ_{Q2} = Average arrival rate for $Q_2 = 0.4\lambda 1 + \lambda 2$
λ_{Q3} = Average arrival rate for $Q_3 = 0.4\lambda 1$
λ_{Q4} = Average arrival rate for $Q_4 = 0.84\lambda 1 + \lambda 2$

If we assume that each of the queues - Q_1, Q_2, Q_3 and Q_4 - are of type M/M/m/∞, then Burke's Theorem and the results quoted earlier for the random splitting and superposition of Poisson flows, indicate that each of the above arrival processes (to each of the queues) will also be Poisson. Consider Q_i where $i=1, 2, 3$ or 4. For this queue, standard M/M/m/∞ results may now be applied with λ_{Qi} as the arrival rate to find $p_{Qi}(n_i)$ as the probability of there being n_i customers in the system at Q_i - this would of course also require knowledge of the mean service at this queue. This computation may be done for each of these queues to find the state probability distribution for each queue. Note that this procedure may be generalized for any queueing network as long as there is no feedback and may be used for any type of M/M/m/∞ queue. (The presence of feedback will introduce dependencies between the output and input processes of the queue and therefore these statements cannot be directly made. However, it turns out that, even in such cases, this approach is still valid.) In this case, we can also obtain (using Jackson's Theorem discussed subsequently) that the joint state distribution for the system overall will be given by the *product of the individual state probabilities* as

$$P(n_1, n_2, n_3, n_4) = p_{Q1}(n_1)p_{Q2}(n_2)p_{Q3}(n_3)p_{Q4}(n_4) \tag{5.1}$$

Jackson's theorem is also the one which shows that the approach given above may be used to give the same kind of *product form solutions*, even when there is feedback present in the system. This theorem is given next.

5.3.1 Jackson's Theorem

Jackson's Theorem is applicable in a *Jackson Network*. This is defined to be an arbitrary network of M/M/m/∞ queueing nodes where jobs arrive in a Poisson stream from outside (to one or more nodes) and are transferred probabilistically from one node to another until their eventual departure from the system. The departures can also occur from one or more queues. (Note that the system is therefore an *open queueing network*.) The M/M/m/∞ queueing nodes are sometimes also referred to as *Jackson Servers*.

Consider a Jackson Network with K such queues and let Λ_i be the average arrival rate from a Poisson process from outside entering the network to the ith queue Q_i. Note that $\Lambda_i=0$ implies that there is no arrival from outside to Q_i. It should also be noted that we are considering an *Open Network* here (*Closed Networks* are considered later) and that for such networks, at least one such $\Lambda_i>0$ must exist for at least one of the queues. A customer served at Q_i is routed to Q_j with routing probability p_{ij} or exits the network with probability $\left[1-\sum_{j=1}^{K} p_{ij}\right]$. Let λ_j be the total average flow rate of arrivals entering Q_j. This may be found by applying the flow balance conditions given in Eq. (5.2) and solving the set of simultaneous equations for the mean arrival rates of jobs $\{\lambda_j\}$ $j=1,\ldots\ldots,$ K to each of the K queues in the network.

$$\lambda_j = \Lambda_j + \sum_{i=1}^{K} \lambda_i p_{ij} \quad \text{for } j = 1,2,\ldots\ldots.K \tag{5.2}$$

For an *Open Network* as being considered here, the K equations given by the above may be solved to obtain the average arrival rate for each of the K queues in the system. (Note that this cannot be done for a *Closed Network*, which is one reason why we consider such networks separately later.) It should also be noted that in the above, we have not restricted the routing probabilities in any manner, which is why they may be such that there is *feedback* present in the network. The set of flow balance equations given in Eq. (5.2) may also be written in a matrix form as

$$[\widetilde{I} - \widetilde{P}^T]\widetilde{\lambda} = \widetilde{\Lambda} \tag{5.3}$$

where

$$\widetilde{P} = [p_{ij}]$$
$$\widetilde{\lambda}^T = [\lambda_1, \ldots\ldots\ldots, \lambda_K]$$
$$\widetilde{\Lambda}^T = [\Lambda_1, \ldots\ldots, \Lambda_K]$$

For a network of this type with M/M/m/∞ queues (i.e. Jackson Servers) at each node, *Jackson's Theorem* states that provided the arrival rate at each queue is such that equilibrium exists, the probability of the overall system state ($n_1, \ldots\ldots, n_K$) will be given as

$$P(\widetilde{n}) = P(n_1, \ldots\ldots, n_K) = \prod_{j=1}^{K} p_j(n_j) \tag{5.4}$$

Here $p_j(n_j)$=P{n_j customers in Q_j}may be found by considering the M/M/m/∞ queue at node j in isolation with its total average arrival rate λ_j, its mean service time $1/\mu_j$ and the corresponding results for the steady state M/M/m/∞ queue. Note that the individual queues, and hence the whole network, will be in equilibrium only if the load $\rho_j = \lambda_j / \mu_j < m_j$ for Q_j, $j=1,2,\ldots,K$, (i.e. for each queue in the network) where m_j is the number of servers at Q_j.

Note that there are some important implications of the statement of Jackson's Theorem as it is given above. [The theorem may be generalized further - such generalizations are mentioned subsequently.] One is that once the average flow rates are determined, the queues may be considered in isolation even when the network is such that there is feedback. The second is that the states of the individual queues behave as if they are independent of each other and therefore the joint probability of the system states will be given as the product of the state probabilities of the individual queues. It should also be noted that the flows entering the individual queues behave as if they are Poisson even when they may not actually be Poisson - such as when the system has feedback. To see the importance of this last statement, consider the simple feedback situation (i.e. immediate feedback) illustrated in Figure 5.7.

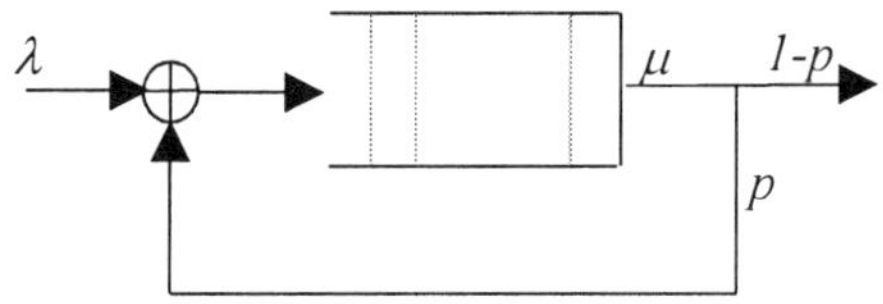

Figure 5.7. Immediate Feedback to a Queue

For the example shown in Figure 5.7, consider the case when $(1-p)<<1$ and $\mu>>\lambda$. In this case, even if the external arrival process is Poisson, the arrival process actually entering the queue will not be Poisson. As a matter of fact, it can be easily seen that because of the feedback, the arrivals to the queue will tend to be in bursts of random size, which will be triggered by the arrival of a customer from outside. This illustrates the fact that the presence of feedback will cause dependencies to arise and will make the arrival process to a queue non-Poisson in nature. The importance of Jackson's Theorem is that it states that even in such a situation, the individual queues may be analysed as if their inputs are Poisson processes and that the overall system state will behave as if the individual queues are independent of each other. Using Jackson's Theorem, the queueing network example given above may be solved for the joint state probability of the system. Some other examples, which incorporate feedback, are given subsequently.

Network Performance Parameters

Once the actual flow rate to each queue in the network has been obtained, Jackson's Theorem may be used to first obtain the state distribution of each queue in isolation and the overall state distribution as the product of the individual distributions. Using these flows and the state distributions, the following parameters of interest are typically calculated.

$$Total\ Throughput = \lambda = \sum_{j=1}^{K} \Lambda_j \tag{5.5}$$

This follows from stability considerations, as the flow in of jobs entering the system must be equal to the flow out of jobs leaving the system, when the system is stable. The average traffic load to each queue may be found as

$$Average\ traffic\ load\ at\ node\ j\ (i.e.\ Q_j) = \rho_j = \frac{\lambda_j}{\mu_j} \tag{5.6}$$

$$Visit\ count\ to\ node\ j = V_j = \frac{\lambda_j}{\lambda} \tag{5.7}$$

The visit count to node j, is a measure of the average number of visits a job makes to the queue Q_j once it enters the queueing network. These may also be found directly by solving the K linear equations of Eq. (5.8).

$$V_j = \frac{\Lambda_j}{\lambda} + \sum_{i=1}^{K} V_i p_{ij} \tag{5.8}$$

It is easily seen that Eq. (5.8) may be directly obtained by normalizing the flow balance equations of Eqs. (5.2) or (5.3) by λ. This is sometimes preferred but the two equations are really the same and if Eq. (5.8) is used to solve for the visit counts then the actual flow to the queues may be obtained using $\lambda_j = \lambda V_j$ for $j=1,2 \ldots\ldots,K$.

$$Average\ number\ of\ jobs\ at\ node\ j = N_j = \sum_{k=0}^{\infty} k p_j(k) \tag{5.9}$$

$$Average\ number\ of\ jobs\ in\ system = N = \sum_{j=1}^{K} N_j \tag{5.10}$$

Note that Eq. (5.9) directly follows from the definition of the state probability for each of the queues, Q_j $j=1,2,.....,K$ and (5.10) is obtained by summing the mean number in each queue in the network. Applying Little's result to Q_j with the average arrival rate λ_j (i.e. $W_j = N_j/\lambda_j$) will then give the total mean delay W_j (queueing delay and service time) that a job will experience every time it visits Q_j. Note that a job enters Q_j for V_j times, every time it enters the system. Therefore, the total time spent by a job in Q_j, every time it enters the system will be $V_jN_j/\lambda_j = N_j/\lambda$. One can also define the performance parameter *mean sojourn time* as the mean total time spent in the system by an arriving job before it leaves the network. This may be found either as the sum of the total times that the job will spend in each of the K queues in the network or one may apply Little's result directly to the overall network and obtain this as N/λ. Both these definitions are given in Eq. (5.11).

$$\textit{Mean Sojourn Time} = W = \frac{N}{\lambda} = \sum_{j=1}^{K} \frac{N_j}{\lambda} \tag{5.11}$$

Actually, the *Product Form* expression for the state probabilities of a queueing network *(Closed* or *Open*, even though we have only mentioned Jackson's Theorem for open networks above*)* holds for any network where the local balance conditions are satisfied. Specifically, open or closed networks with the following types of queues will all have a product form solution -

1. FCFS queues with exponentially distributed service times
2. LCFS pre-emptive resume queues with service times having a Coxian distribution
3. Processor Sharing (PS) queues with service times having a Coxian distribution
4. Infinite Server (IS) queues with service times having a Coxian distribution.

A *Coxian Distribution* is one where the L.T.. of the service time's probability density function is of the following form.

$$L_B(s) = \gamma_1 + \sum_{i=1}^{r} \beta_1 \beta_2 .....\beta_i \gamma_{i+1} \prod_{j=1}^{i} \frac{\mu_j}{s+\mu_j}$$

with $\beta_i = 1 - \gamma_i$ for $1 \leq i \leq r$ and $\gamma_{r+1} = 1$

The *Exponential Distribution* may be observed to be a special case of a *Coxian Distribution*. It should also be noted that for queues of the type mentioned above, an arrival process that is Poisson in nature, would lead to a departure process that is also Poisson in nature. This result is once again useful because a Poisson process may be randomly split or different Poisson processes may be combined without losing the Poisson property.

We consider next some actual examples of open networks which may be analysed using the approach given in this section.

5.3.2 Some Examples of Jackson Networks

Example 1

Consider the queueing network shown in Figure 5.8 which is an open queueing network with feedback.

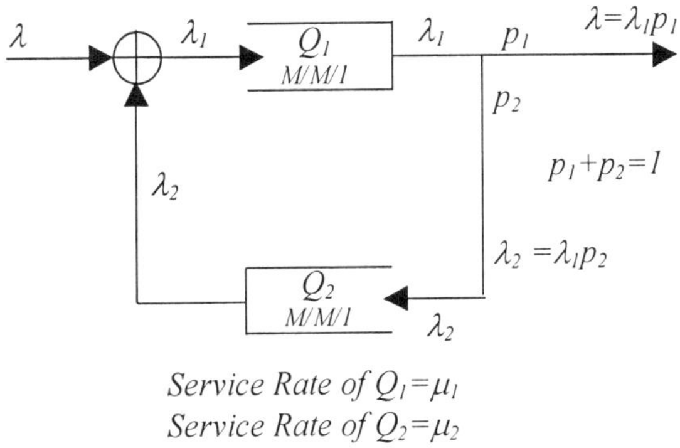

Figure 5.8. Open Queueing Network of Example 1

For this simple system, we can immediately observe from *Flow Balance* that

$$\lambda_1 = \frac{\lambda}{p_1} \quad \text{and} \quad \lambda_2 = \frac{\lambda(1-p_1)}{p_1}$$

Therefore, $\rho_1 = \dfrac{\lambda}{\mu_1 p_1}$ and $\rho_2 = \dfrac{\lambda(1-p_1)}{\mu_2 p_1}$

Using the results for M/M/1 queues, one can therefore get

System State Probability: $P(n_1, n_2) = \rho_1^{n1}(1-\rho_1)\rho_2^{n2}(1-\rho_2)$

Mean Number in the Queues: $N_1 = \dfrac{\rho_1}{1-\rho_1}$ and $N_2 = \dfrac{\rho_2}{1-\rho_2}$

Therefore $N = N_1 + N_2$ will be the total mean number of users in the overall system. We can also apply Little's Formula to get the total time spent by a job entering the system to be

$$W = \frac{N}{\lambda} = \frac{\rho_1}{\lambda(1-\rho_1)} + \frac{\rho_2}{\lambda(1-\rho_2)}$$

Example 2

Consider the network of Figure 5.9, where arrivals can enter the network at two points with rates *r* and *2r*. The queues Q_1 and Q_2 have service rates 2μ and μ, respectively

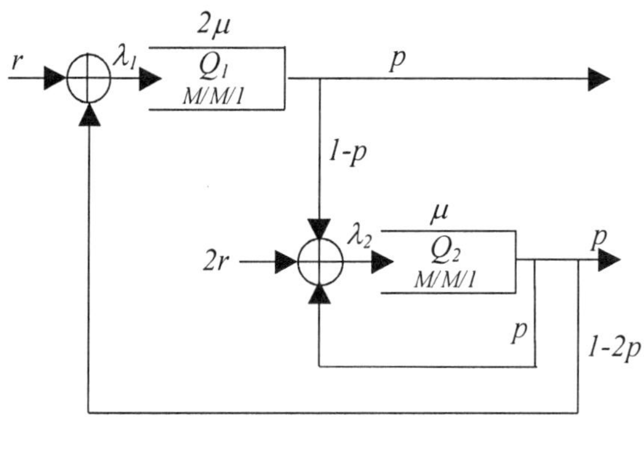

Figure 5.9. Open Queueing Network of Example 2

From flow balance conditions, we get

$$\lambda_1 = r + \lambda_2(1-2p)$$
$$\lambda_2 = 2r + \lambda_1(1-p) + \lambda_2 p$$

Solving these we get

$$\lambda_1 = \frac{r(3-5p)}{2p(1-p)} \qquad \lambda_2 = \frac{r(3-p)}{2p(1-p)}$$

$$\rho_1 = \frac{r(3-5p)}{4\mu p(1-p)} \qquad \rho_2 = \frac{r(3-p)}{2\mu p(1-p)}$$

The joint state probability of the two queues in the system will then be given by

$$P(n_1, n_2) = (1-\rho_1)(1-\rho_2)\rho_1^{n1}\rho_2^{n2}$$

The average visit counts to the two queues will be given by

$$V_1 = \frac{3-5p}{6p(1-p)} \qquad V_2 = \frac{3-p}{6p(1-p)}$$

Using, M/M/1 results, we can find the mean number in the two queues to be

$$N_1 = \frac{\rho_1}{1-\rho_1} \qquad N_2 = \frac{\rho_2}{1-\rho_2}$$

The mean of the total number of jobs in the system will be

$$N = \frac{\rho_1}{1-\rho_1} + \frac{\rho_2}{1-\rho_2}$$

The mean sojourn time W of a job entering the system (at either of the two entry points) may then be found to be $N/3r$. This follows from the application of Little's result to the value of N obtained above.

Example 3

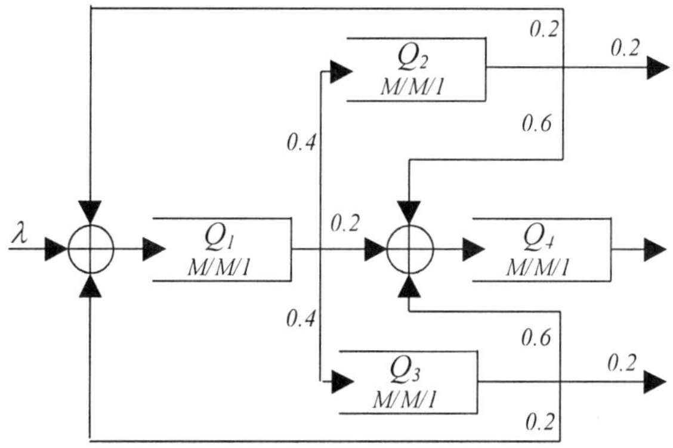

Figure 5.10. Open Queueing Network of Example 3

Consider the open network of single-server M/M/1 type queues shown in Figure 5.10, where the mean service times at the queues Q_1, Q_2,, Q_3 and Q_4 are specified to be $1/\mu$, $2/\mu$, $2/\mu$ and $2/\mu$ seconds, respectively. Let λ be the average external arrival rate (Poisson) to Q_1. Apart from the usual performance parameters as considered in our earlier examples, in this case, we would also like to find the maximum value of λ under which the network will operate in a stable fashion. For this, we identify the first queue that becomes unstable as λ increases and the mean number in each of the other queues when the first one becomes unstable. The flow equations for the network are

$$0.2\lambda_2 + 0.2\lambda_3 + \lambda = \lambda_1$$
$$\lambda_2 = \lambda_3 = 0.4\lambda_1$$
$$\lambda_4 = 0.2\lambda_1 + 0.6\lambda_2 + 0.6\lambda_3$$

Solving these, we get the solution for the mean arrival rates to each queue to be

$$\tilde{\lambda} = (\lambda_1, \lambda_2, \lambda_3, \lambda_4) = (1.1905\lambda, 0.4762\lambda, 0.4762\lambda, 0.8095\lambda)$$

and the traffic is

$$\tilde{\rho} = (\rho_1, \rho_2, \rho_3, \rho_4) = (1.1905\rho_0, 0.9524\rho_0, 0.9524\rho_0, 1.6190\rho_0)$$

with $\rho_0 = \lambda/\mu$.

Following our usual approach, we obtain the system state probability as

$$P(n_1, n_2, n_3, n_4) = (1-\rho_1)(1-\rho_2)^2(1-\rho_4)\rho_1^{n_1}\rho_2^{n_2+n_3}\rho_4^{n_4} \quad \text{for } 0 \le n_1, n_2, n_3, n_4 \le \infty$$

and $N_1 = \dfrac{\rho_1}{1-\rho_1}$ $\quad N_2 = N_3 = \dfrac{\rho_2}{1-\rho_2}$ $\quad N_4 = \dfrac{\rho_4}{1-\rho_4}$

Note that ρ_4 is the largest queue traffic occurring at $Q4$. Therefore as the offered traffic λ increases, this is the queue that we expect to become unstable first. The condition $\rho_4 < 1$ is then the required stability condition for the network. This, therefore, implies that $\rho_0 < 0.6176$ or $\lambda < 0.6176\mu$ for the queueing network to be stable.

The stability limit of $\rho_4 < 1$ also implies that at the point when the network is about to become unstable, the traffic at the queues Q_1, Q_2 and Q_3 will be $\rho_1 < 0.7352$ and $\rho_2 = \rho_3 < 0.5882$, respectively. At this stability limit, the number in each of the queues Q_1, Q_2, Q_3 and Q_4 will be $N_1 = 2.7764$, $N_2 = N_3 = 1.4284$ and $N_4 \to \infty$.

5.4 Extensions to Jackson's Theorem for Other Open Networks

The statement of Jackson's Theorem can actually be generalized to cover more complex queueing networks than the one discussed earlier in Section 5.4. The statements of Jackson's Theorem for two such extensions are summarized next.

5.4.1 Jackson's Theorem with State Dependent Service Rates at the Queueing Nodes

For this, assume that the service times at Q_j are exponentially distributed with mean $1/\mu_j(m)$ when there are m customers in Q_j just before the departure of a customer (Note that m would include the departing customer.) Unlike the network considered in Sec. 5.3, here the service rate $\mu_j(m)$ is actually dependent on the number of customers currently in the queue. This approach may be used to model multi-server M/M/- type queues in a very general fashion.

Consider a queueing network of K queues formed with queues like Q_j, as described above, where the other assumptions remain the same as before in Section 5.3. Let the load $\rho_j(m)$ at Q_j when there are m customers in that queue, be defined as

$$\rho_j(m) = \frac{\lambda_j}{\mu_j(m)} \qquad \text{for } j=1, \ldots, K \text{ and } m=1, 2, \ldots\ldots$$

where the λ_j's are obtained by solving the flow balance equations

$$\lambda_j = \Lambda_j + \sum_{i=1}^{K} \lambda_i p_{ij} \qquad \text{for } j=1, \ldots, K$$

as before. We also define

$$\hat{P}_j(n_j) = 1 \qquad n_j = 0$$

$$= \prod_{k=1}^{n_j} \rho_j(k) \qquad n_j > 0$$

Then Jackson's Theorem for this *network of queues with state dependent service rates* states that the system state probability will be given by

$$P(\tilde{n}) = P(n_1, n_2, \ldots\ldots, n_K) = \frac{1}{G} \prod_{k=1}^{K} \hat{P}_k(n_k)$$

with *Normalisation Constant* $G = \sum_{n_1=0}^{\infty} \cdots\cdots \sum_{n_K=0}^{\infty} \prod_{i=1}^{K} \hat{P}_i(n_i)$, $0<G<\infty$.

It is important to note that the system's state probability distribution is still in the form of continued products and hence this solution is also classifiable as a product-form solution. For any given queue Q_j, the service rate when the queue is in state m can be arbitrarily specified and this specification may be done differently for different queues. This is a very powerful feature of this extension to Jackson's theorem as it not only allows the individual queues to be modelled in a variety of different ways, but also allows queues of very different types to be mixed together in the same queueing network. As a simple application of this concept, we can consider queueing networks made up of multi-server queues. Consider a particular queue with s servers. We can then consider the service rate $\mu(m)$ for that queue to be $m\mu$ for $m \leq s$, and $s\mu$ for $m > s$. Here we assume μ to be the service rate of one server and the service times to be exponentially distributed with mean $1/\mu$.. Other more complicated service scenarios may also be modelled in this fashion. Note that, as assumed in Section 5.3, we still require the individual service times to be exponentially distributed and the queues to be ones with infinite buffers.

5.4.2 Queueing Networks with Multiple Customer Classes

Jackson's Theorem is also applicable to networks serving multiple classes of customers, provided that at each queue the service time distribution is the same for all customer classes. Consider such a queueing network with C customer classes (i.e. $c=1, \ldots\ldots, C$) with

(a) average rates $\Lambda_j(c)$ for Poisson arrivals of class c from outside to the network at node j

(b) routing probabilities $p_{ij}(c)$ giving the (transition) probability that a job of class c completing service at node i, goes to node j next

Assume that the mean service time (exponentially distributed) at Q_j when there are m users in that queue, is given by $1/\mu_j(m)$ as in Section 5.4.1 earlier. Note that this is also *state dependent* as in the previous case but is now required to be the same for all the C classes. We define the vector

$$z_j = (c_1, \ldots\ldots\ldots, c_{n_j})$$

where n_j is the number of customers in Q_j and c_i is the class of the customer in the ith position in Q_j. Note that the vector z_j essentially gives the composition of Q_j at a given instant of time. The *state of the network* at a given time instant may then be represented by the vector $\tilde{z}$, where

$$\tilde{z} = (z_1, \ldots\ldots\ldots, z_K)$$

We can solve the flow balance equations for each type of job $c=1,\ldots,C$ to obtain the arrival rate of class c customers to Q_j as $\lambda_j(c)$ $j=1,\ldots K$ and $c=1,\ldots,C$. Using this and the service rate model for each of the queues, we can define the following load parameter.

$$\hat{\rho}_j(c,m) = \frac{\lambda_j(c)}{\mu_j(m)}$$

We also define

$$\hat{P}_j(z_j) = 1 \qquad\qquad n_j = 0$$

$$= \prod_{k=1}^{n_j} \hat{\rho}_j(c_k, k) \qquad n_j > 0$$

and $G = \sum_{(z_1,......,z_K)} \prod_{j=1}^{K} \hat{P}_j(z_j)$ as the *normalisation constant.*

Jackson's Theorem for this type of queueing network then states that the state probability of this system will be given by the following product form expression.

$$\hat{P}(\tilde{z}) = \frac{1}{G} \prod_{j=1}^{K} \hat{P}_j(z_j)$$

5.5 Closed Queueing Networks

We showed an example of a Closed Queueing Network earlier in Figure 5.2. Here jobs would continually circulate around the network as shown, with probabilistic routing between the nodes. There are no arrivals from outside nor are there any departures from the system to the outside world. The system is started with a certain number of jobs and these jobs continually circulate in the network moving from one node to another on completing service at the former. The destination node, to which the job moves on completing service at the earlier queue, is decided based on the probabilistic routing specified by the routing probabilities.

Consider a closed network of this type with K queues - Q_1, ..., Q_K. Assume that there are M jobs of the same class circulating in the network. (As mentioned in the case of open networks, it is also possible to have multiple classes of customers for closed networks as well. The formulation of the problem and the results may be extended to cover this case which will be somewhat more complicated but solvable in a similar fashion.) Let p_{ij} be the routing probability from Q_i to Q_j - i.e. a job finishing service at Q_i will be routed to Q_j with this probability. Note that if Q_i to Q_j is the only transition possible, then $p_{ij}=1$. Since we are considering a closed network, jobs must be routed to one of the queues in the network and the sum of the transition probabilities from Q_i to any other Q_j in the network (including Q_i itself) must add up to unity. This implies that we have

$$\sum_{\forall j} p_{ij} = 1 \quad \forall i$$

in a closed network.

Let λ_j be the total average arrival rate to Q_j $j=1,\dots\dots, K$ where application of *flow balance* would lead to the following K equations

$$\lambda_j = \sum_{i=1}^{K} \lambda_i p_{ij} \qquad j=1, \dots\dots, K \tag{5.12}$$

Note that these equations are *not independent* (since there are no external arrivals to the network) and hence (unlike the *Open Network* case) they cannot be solved to uniquely find the λ_js for the K queues, $j=1,\dots, K$. However, using any of the K-1 equations in the above, we can find the λ_j's up to a multiplicative constant. In order to do this, assume that $\alpha(M)$ is an (unknown) scalar quantity and let $\{\lambda_j^*\}$ $j=1,\dots,K$ be a particular solution of Eq. (5.12) such that the *true average arrival rates* $\{\lambda_j(M)\}$ $j=1,\dots,K$ are given by

$$\lambda_j(M) = \alpha(M)\lambda_j^* \quad j=1,\dots\dots K \tag{5.13}$$

Note that both $\alpha(M)$ and $\{\lambda_j(M)\}$ are functions of the population size M (i.e. the total number of jobs circulating in the system) though $\{\lambda_j^*\}$ $j=1,\dots,K$ are independent of M.

An alternate approach which is equivalent is to choose any one queue in the network (say Q_1) as the reference queue and assume that $\lambda_1^*=\alpha$, where any value of α may be chosen. (One usually chooses something which is convenient such as $\alpha=\mu_1$ so that $\rho_1=\lambda_1^*/\mu_1$ becomes equal to 1, simplifying calculations.) We can then use the flow balance equations given by Eq. (5.13) to obtain $(\lambda_2^*, \lambda_3^*, \dots\dots, \lambda_K^*)$ in terms of α. Note that $(\lambda_1^*, \lambda_2^*, \dots\dots, \lambda_K^*)$ are referred to as the *Relative Throughputs* with respect to the reference queue - which, in this case, was selected to be Q_1 (any other queue may also be chosen as the reference queue).

Let $\mu_j(m)$ be the (state dependent) service rate at Q_j (where the service times are exponentially distributed) when Q_j is in state m. Note that the service times for a particular job at a queue are still assumed to be exponentially distributed even though the service rates at the queue may vary depending on the state of the queue. Using any particular solution $\{\lambda_j^*\}$ of Eq. (5.12), we can then define

$$u_j(m) = \frac{\lambda_j^*}{\mu_j(m)} \qquad j=1, \ldots, K; \quad m=1, \ldots, M \tag{5.14}$$

as the relative utilization of Q_j with respect to the relative utilization of the reference queue. Let

$$\begin{aligned} \hat{P}_j(n_j) &= 1 & n_j &= 0 \\ &= u_j(1)u_j(2)\ldots\ldots u_j(n_j) & n_j &\geq 1 \end{aligned} \tag{5.15}$$

and the corresponding normalisation constant $G(M)$ for a population M is defined to be -

$$G(M) = \sum_{n_1+\ldots+n_K=M} \hat{P}_1(n_1)\hat{P}_2(n_2)\ldots\ldots\hat{P}_K(n_K) \tag{5.16}$$

We can now state *Jackson's Theorem for Closed Networks* using these quantities.

5.5.1 Jackson's Theorem for Closed Networks

Jackson's Theorem for Closed Networks states that for all states $n_1, \ldots, n_K$ such that $n_1+n_2+\ldots+n_K = M$, the probability of the state vector $(n_1, n_2, \ldots, n_K)$ will be given by the following *product-form* expression.

$$P(\tilde{n}) = P(n_1, n_2, \ldots, n_K) = \frac{1}{G(M)} \prod_{i=1}^{K} \hat{P}_i(n_i) \tag{5.17}$$

Note that the expression is indeed in a *product-form* (i.e. continued product of terms) as we would expect for a *Jackson Network*. It should be noted that the only restriction in the above result is that the individual service time of a job at any of the queues is exponential in nature. We also require that the service rate for Q_j with m jobs in the queue may be expressed in the form $\mu_j(m)$ as described above and that the actual service times of individual jobs are exponentially distributed. Apart from these restrictions, the theorem is quite general and can be applied to handle queues where the service rate varies as a function of the system state. We illustrate this with a simple example next.

Example

Consider the closed network with two queues Q_1 and Q_2 shown in Figure 5.11 with M jobs in the system. Both Q_1 and Q_2 are single server queues with service rates μ_1 and μ_2, respectively, where the service times are exponentially distributed. The routing probabilities are p, $1-p$, q, $1-q$ as shown in Figure 5.11. Note that the M jobs will continually circulate in the network moving from one queue to another as per the routing probabilities indicated. At any given instant of time, some of these M jobs will be in one queue while the remaining jobs will be in the other queue.

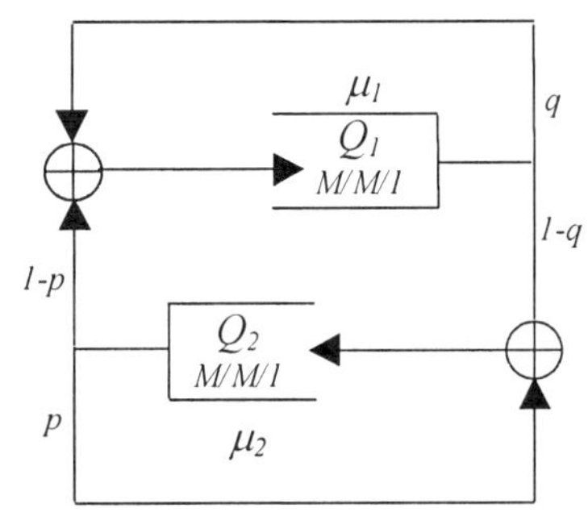

Figure 5.11. A Closed Network with M Jobs

We choose the particular solution λ_1^* for Q_1 to be $\lambda_1^* = \mu_1$. Then from the flow balance equations we will get

$$\lambda_2^* = \lambda_1^* \frac{1-q}{1-p} = \mu_1 \frac{1-q}{1-p}$$

and, therefore

$$u_1 = 1, \quad u_2 = \frac{1-q}{1-p}\frac{\mu_1}{\mu_2}$$

The Steady State Distribution of the M users in the system (i.e. n_1 in Q_1 and n_2 in Q_2, such that $n_1+n_2=M$) will then be given by the product form solution to be

$$P(n_1, n_2) = P(M-n, n) = \frac{u_2^n}{G(M)}$$

where $G(M)$ is the normalisation constant (for M jobs circulating in the system). This will be given by

$$G(M) = \sum_{n=0}^{M} u_2^n = \frac{1-u_2^{M+1}}{1-u_2}$$

This state probability distribution may now be used to find the various performance parameters of the individual queues and of the overall system. For example, the probabilities of the queues to be busy, will be given by

$$\text{P}\{Q_1 \text{ is busy}\} = 1\text{-P}(0, M) = 1 - \frac{u_2^M}{G(M)} = \frac{G(M-1)}{G(M)}$$

$$\text{P}\{Q_2 \text{ is busy}\} = 1\text{-P}(M, 0) = 1 - \frac{1}{G(M)} = u_2 \frac{G(M-1)}{G(M)}$$

Visit Ratios in a Closed Queueing Network

The *visit ratio* V_i of the ith queue Q_i in the queueing network is defined as the mean number of times Q_i is visited by a job/customer for every visit it makes to a given reference queue, say Q_1. (Note that this definition is very similar to the definition of visit ratios in the case of open networks)

If Q_1 is chosen as the reference queue, we will then have

$$V_i = \frac{\lambda_i^*}{\lambda_1^*} \qquad i=1, 2, \ldots, K \tag{5.18}$$

We can also obtain the visit ratios directly by solving for them in the same way as solving for the relative throughputs ($\lambda_1^*, \ldots \lambda_K^*$). In vector form, this implies solving for the vector

$$\widetilde{V} = (V_1, \ldots\ldots, V_K)$$

using the following vector equation with $V_1 = 1$

$$\widetilde{V} \cdot \widetilde{P} = \widetilde{V} \tag{5.19}$$

where $\widetilde{P} = [p_{ij}]$ is the matrix of the routing probabilities p_{ij} of going from Qi to Q_j.

Relative Utilizations of Queues in a Closed Queueing Network

Note that the *relative utilization* u_i of Q_i is defined as above to be

$$u_i = \frac{\lambda_i^*}{\mu_i} \tag{5.20}$$

This definition implies that the relative utilization of a queue is being defined with reference to the relative utilization of the reference queue. As mentioned earlier, a proper choice of the relative throughput of the reference queue may be done such that its relative utilization becomes unity. This is generally desirable, as it would then simplify subsequent calculations to some extent.

5.5.2 Jackson's Theorem for Closed Networks of Multi-Server Queues

We consider here the specific case of a *Closed Network of Multi-Server Queues* where there are M jobs circulating in K queues $Q_1, \ldots\ldots, Q_K$ and where Q_j has s_j servers such that the service rate in Q_j when it has m jobs/customers is $\mu_i(m) = min(m\mu, s\mu)$ with exponentially distributed service times. For this network, an alternative statement of the *product-form solution* (given earlier in Eqs. (5.16) and (5.17)) for the probability of the system being in state $\widetilde{n} = (n_1, \ldots\ldots, n_K)$ is

$$P(\widetilde{n}) = P(n_1, \ldots\ldots, n_K) = \frac{1}{G(M)}\left[\prod_{i=1}^{K} \frac{u_i^{n_i}}{\beta_i(n_i)}\right] \tag{5.21}$$

for all $\tilde{n} = (n_1,.....,n_K)$ such that $n_1+n_2+......+n_K = M$. Here $\beta_i(n_i)$ is given by

$$\begin{aligned} \beta_i &= n_i! & n_i \le s_i \\ &= s_i!(s_i)^{(n_i - s_i)} & n_i > s_i \end{aligned} \tag{5.22}$$

and $G(M)$ is the Normalisation Constant given by

$$G(M) = \sum_{n_1+....+n_K=M} \left(\prod_{i=1}^{K} \frac{u_i^{n_i}}{\beta_i(n_i)} \right) \tag{5.23}$$

For the analysis of closed networks of multi-server queues of this type, using Eqs. (5.21)-(5.23) is generally somewhat simpler than using the original product form expressions of Eqs. (5.16) and (5.17).

In the solution of *Closed Queueing Networks*, the primary difficulty is usually in the calculation of the *Normalisation Constant G(M)*. This computation becomes increasingly difficult to do in a direct way when the number of queues and/or the population of jobs in the system become large, i.e. by evaluating each component term in Eqs. (5.16) or (5.23) and summing all of them individually. This problem can be avoided in one of two ways.

If the mean system performance parameters are the only ones that are really desired (and not the state probabilities), then these may be directly computed using the *Mean Value Algorithm* which is based on Reiser and Lavenberg's *Mean Value Theorem* [ReL80]. This is a computationally efficient way of calculating the mean number in each queue and the mean time spent in the queue. Other mean performance measures may also be subsequently computed from these parameters.

The other method suggested is a *Convolution Algorithm* [Mol89], [LZS84] for calculating the Normalisation Constant $G(M)$. This method also provides $G(M-n)$, $n=0,...,$ $M-1$ in the intermediate stages of the calculation Note that $G(M-n)$ is the normalisation constant for the same network when there are $M-n$ jobs in the system, instead of M. The normalisation constant $G(M)$ may be used to get the state probabilities of the system as mentioned earlier. In some cases, the normalisation constants $G(1),.........,G(M)$ may also be used by themselves to easily obtain various system performance measures in a direct and simple fashion. Note that both these algorithms can only be applied to queues where the individual service times are exponentially distributed.

For a closed network of K *single server queues* with M circulating jobs, it is also possible to simplify Eqs. (5.16)-(5.20) to get the following results

$$P(\tilde{n}) = P(n_1, \ldots\ldots, n_K) = \frac{1}{G(M)}\left[\prod_{i=1}^{K} u_i^{n_i}\right] \text{ for } n_1 + \ldots\ldots + n_K = M \qquad (5.24)$$

with *G(M)* as the Normalisation Constant given by

$$G(M) = \sum_{n_1+\ldots+n_K=M} \left(\prod_{i=1}^{K} u_i^{n_i}\right) \qquad (5.25)$$

For this, consider the case where we want to compute $P\{n_j \geq n\}$, i.e. the probability that Q_j contains n or more customers. In this case, we would need to sum $P(\tilde{n})$ over all states $\tilde{n}$ satisfying this condition. This may be shown to be

$$P\{n_j \geq n\} = u_j^n \frac{G(M-n)}{G(M)} \qquad j=1, \ldots\ldots, K \qquad (5.26)$$

The actual utilization of a single-server queue is merely the probability that the queue is not empty. Therefore the actual utilization ρ_j of Q_j will be -

$$\rho_j = u_j \frac{G(M-1)}{G(M)} \qquad j=1, \ldots\ldots, K \qquad (5.27)$$

and its actual throughput λ_j will be -

$$\lambda_j = \mu_j \rho_j = \mu_j u_j \frac{G(M-1)}{G(M)} \quad j=1, \ldots\ldots, K \qquad (5.28)$$

for each queue Q_j in the network, $j=1, \ldots\ldots, K$. Moreover, using the general result that

$$E\{n\} = \sum_{n=1}^{\infty} nP\{n\} = \sum_{n=1}^{\infty}\sum_{k=n}^{\infty} P\{k\} = \sum_{n=1}^{\infty} P\{k \geq n\} \qquad (5.29)$$

we can derive the mean number $E\{n_j\}$ in Q_j to be

$$E\{n_j\} = \sum_{m=1}^{M} u_j^m \frac{G(M-m)}{G(M)} \quad j=1, \ldots\ldots, K \tag{5.30}$$

Note that the results given in (5.26)-(5.30) are really useful if the normalisation constants for the queueing network of single server queues are known for $m=1,\ldots\ldots,M$ (where M is the actual population of jobs in the system). This, for example would be the case if the *convolution method* is being used to solve the network. In that case, these equations may be used to quickly obtain the performance results for the individual queues in the network. This method is described in the next section.

5.6 Convolution Algorithm for Finding the Normalisation Constant for a Closed Queueing Network

We present here the convolution method for finding the *normalisation* constant $G(M)$ for a closed network of K queue with M jobs circulating in the network. The jobs are considered to have exponentially distributed service times. We consider separately the two cases (a) where all the queues are single-server queues and (b) where one or more queues may have more than one server or when one or more queues have state-dependent service rates. The algorithm given for the latter may be used for the network of single server queues as well. However, the former algorithm is more simply stated and easier to use for actual calculations.

5.6.1 Network of Single Server Queues

Define the variable $g(n, k)$ as the normalisation constant when there are n jobs circulating in a network with k queues where the individual queues have the same relative utilizations as in the original network. This is given by

$$g(n,k) = \sum_{n_1+\ldots+n_K=n} \left(\prod_{i=1}^{k} u_i^{n_i} \right) \tag{5.31}$$

This may be expressed equivalently as

$$g(n,k) = u_1^{\,n} + u_1^{\,n-1}u_2 + \ldots\ldots\ldots + u_1^{\,n-1}u_k + u_1^{\,n-2}u_2^{\,2} + \ldots\ldots + u_k^{\,n} \tag{5.32}$$

We can then write $g(n-1, k)$ and $g(n, k-1)$ as

$$g(n-1,k) = u_1^{\,n-1} + u_1^{\,n-2}u_2 + ...+ u_1^{\,n-2}u_k + u_1^{\,n-3}u_2^{\,2} + ...+ u_k^{\,n-1} \tag{5.33}$$

$$g(n,k-1) = u_1^{\,n} + u_1^{\,n-1}u_2 + ...+ u_1^{\,n-1}u_{k-1} + u_1^{\,n-2}u_2^{\,2} + ...+ u_{k-1}^{\,n} \tag{5.34}$$

This leads to the following recursion for $g(n, k)$

$$g(n.k) = g(n,k-1) + u_k g(n-1,k) \tag{5.35}$$

The recursion of Eq. (5.35) is basically the one needed where we start the recursion with the initial values $g(0,k)=1$ and $g(n,1)= u_l^{\,n}$ for $k=1, ...,K$ and $n=1,......,M$. The required normalisation constant $G(M)$ is then given by

$$G(M) = g(M,K) \tag{5.36}$$

which is obtained at the end of the recursion process. It should be noted that, as mentioned earlier, the normalisation constant when there are L jobs circulating in the network, for $L=1,....,M-1$ is also obtained as intermediate results during the actual process of recursion. These values may then be conveniently used, as in Eqs. (5.25)-(5.30), to find the performance parameters of the network.

5.6.2 Network of Multi-Server Queues or Network of Queues with State Dependent Service Rates

The convolution method may also be conveniently used for finding the normalisation constant for a network with K queues with *state dependent service rates* when there are M jobs circulating in the system. Note that the case of multi-server queues will just be a special case of this.

For this assume that the *state dependent service rate* at Q_i when there are j customers in the queue is given by $\mu_i(j)$. Note that this will be the value of the service rate until just before the departure of the jth customer from Q_i. We can define the relative utilization for Q_i this case to be

$$u_i = \frac{\lambda_i^{\,*}}{\mu_i(1)} \tag{5.37}$$

We define a quantity $A_i(n)$ as follows

$$A_i(n) = \prod_{j=1}^{n} \frac{\mu_i(j)}{\mu_i(1)} \tag{5.38}$$

Note that for the case of multi-server queues (but not for queues with general state-dependent service rates) $A_i(n)$ will be the same as $\beta_i(n)$. Using these, we can now write the recursion for $g(n, k)$ as

$$g(n,k) = \sum_{j=0}^{n} \frac{(u_k)^j}{A_k(j)} g(n-j, k-1) \tag{5.39}$$

with the initial values $g(0, k)$ and $g(n, 1)$ given as -

$$g(0,k) = 1 \quad \text{and} \quad g(n,1) = \frac{(u_1)^n}{A_1(n)} \qquad k=1,\ldots,K \quad n=1,\ldots,M \tag{5.40}$$

As in the earlier case, the recursion ends when $g(M, K)$ has been calculated to provide the required normalisation constant $G(M)$ as

$$G(M) = g(M, K) \tag{5.41}$$

5.7 Mean Value Analysis (MVA) Algorithm for a Closed Queueing Network

This algorithm directly computes the mean performance measures of the network without computing the normalisation constant or evaluating the state probabilities though extensions to compute state probabilities are also possible. Since the performance parameters are directly computed, this algorithm is a very convenient one to use if only these parameters, and not the actual state probabilities, are really required. This method is based on Resire and Lavenberg's *Mean Value Theorem*. This theorem [ReL80] states that *a customer arriving to a queue in a Product Form Network sees the same average number in the queue as an outside observer will see if the network had one less customer.*

We do not consider the proof of this theorem in this text but it is easy to see how it helps in analysing a closed queueing network. We can add jobs/customers to the network, one at a time, to derive the queueing results for the case of $n+1$ jobs in the system from the results obtained earlier when there were n jobs in the system. The total delay seen by a job being added to one such queue will then be the newly added job's mean service time and the total delay seen in that queue when there was one less job in the system. This last statement follows from the statement of the Mean Value Theorem. This approach is easy to see in the recursive MVA algorithm given subsequently, which is started with zero jobs in the system (when all queues will be empty and all queueing delays will be zero) and is continued until M jobs have been added to the network.

We give next the actual statement of the *Mean Value Algorithm (MVA)* that has been devised using this theorem for analysing a queueing network. Note that as expected, the mean performance results are the same for FCFS, LCFS or Processor Sharing (PS) queues. (In a processor-sharing queue, the service resources are fixed but are shared equally by all the customers in the queue.) The Infinite Server (IS) queue is a special queue, which has an infinite number of servers working at the same service rate independent of the number in the queue. However, since there are an infinite number of servers, all the customers present in the IS queue will get served simultaneously. Note this also happens with a PS queue, except that in this case, since the service resources are fixed, the service rate available to the individual jobs in the queue will go down linearly as the number of jobs in the queue goes up. In the MVA algorithms given next, the individual queues may be FCFS, LCFS, PS or IS in nature and these can be mixed in any manner in the queueing network being analysed. Note once again that this algorithm may only be applied if the jobs have service times that are exponentially distributed.

In the following, we consider separately the MVA algorithms for the case of (a) closed queueing networks of single server queues and (b) closed queueing networks of multi-server queues. The algorithm of (a) may be obtained from that of (b). However, it is still useful to state the algorithm for single-server queues separately as it is substantially simpler to understand and use than the more general one for multi-server queues.

We need to introduce the following notation for our statement of the MVA algorithm given subsequently. Note that we are considering the application of this algorithm for a closed queueing network of K queues with M jobs circulating in the network.

$N_j(n)$	Mean number of jobs in Q_j when there are a total of n in the network. This includes the job currently being served at Q_j.
$W_j(n)$	Mean time spent by a job in the queue Q_j when there are n in the network. This is the total delay at Q_j including the service time of the job.
$1/\mu_j$	Mean service time for a job at Q_j Note that this will be the same regardless of the number of jobs/customers in Q_j
V_j	Visit Ratio of Q_j. This is as defined earlier in Eq. (5.18)

5.7.1 Network of Single Server Queues with Single Traffic Class

Initialise for each queue with $N_k(0)=0$ $\quad k=1, 2, \ldots\ldots, K$

Repeat Steps 1, 2, & 3 given next, for $m=1, 2, \ldots\ldots M$. This will increase the number of customers by *1*, in sequence, until the final population of M is reached.

1. For each queue Q_k in the network, $k=1, 2, \ldots.., K$, calculate

$$\begin{aligned} W_k(m) &= \frac{1}{\mu_k} && \textit{for IS queues} \\ &= \frac{N_k(m-1)+1}{\mu_k} && \textit{for FCFS, LCFS, PS queues} \end{aligned} \tag{5.42}$$

2. Using Little's result, set the overall throughput as

$$\lambda = \frac{m}{\sum_{k=1}^{K} W_k(m) V_k} \tag{5.43}$$

3. For each of the queues Q_k in the network, $k=1, 2, \ldots., K$, update $N_k(m)$ as

$$N_k(m) = V_k \lambda W_k(m) \tag{5.44}$$

As mentioned earlier, the above recursion is terminated when m reaches the desired number of jobs M circulating in the network. When that happens, the algorithm will directly provide the following performance results for each queue in the network -

(a) $W_k = W_k(M)$ as the total average time spent by a job in queue Q_k of the network, $k=1, 2, \ldots., K$. From this, the queueing delay W_{qk} at the queue Q_k may be found to be

$$\begin{aligned} W_{qk} &= 0 && \textit{for IS queues} \\ &= W_k - \frac{1}{\mu_k} && \textit{for FCFS, LCFS, PS queues} \end{aligned} \tag{5.45}$$

(b) $N_k=N_k(M)$ as the average total number of jobs in Q_k (including the one in service), $k=1,2, \ldots, K$. From this, or using Little's reult on Eq. (5.45), we can also find the mean number of jobs N_{qk} waiting in queue at Q_k.

(c) The average system throughput λ will be directly available from Eq. (5.43) at the final step when it is calculated for $m=M$. The average throughput λ_k of the queue Q_k in the network, $k=1, 2, \ldots, K$, may be found using

$$\lambda_k = \lambda V_k \qquad k=1,2, \ldots, K \tag{5.46}$$

using the visit ratios obtained earlier using Eqs. (5.18) or (5.19).

5.7.2 Network of Multi-Server Queues with Single Traffic Class

Initialise for each queue with $N_k(0)=0$, $p_k(0,0)=1$, and $p_k(j,0)=0$ for $j=1,\ldots,(c_k-1)$, $k=1,\ldots,K$ where Q_k has c_k servers.

Repeat Steps 1, 2, 3 & 4 given next, for $m=1, 2, \ldots M$. This will increase the number of customers by 1, in sequence, until the final population of M is reached.

1. For each queue Q_k in the network, $k=1, 2, \ldots, K$, calculate

$$\begin{aligned} W_k(m) &= \frac{1}{\mu_k} && IS \\ &= \frac{N_k(m-1)+1}{\mu_k} && Single\ Server\ FCFS, LCFS, PS \\ &= \frac{N_k(m-1)+1+S_k}{\mu_k} && Multiple\ Server\ FCFS \end{aligned} \tag{5.47}$$

with S_k defined as

$$S_k = \sum_{j=1}^{c_k-1}(c_k - j)p_k(j-1, m-1) \tag{5.48}$$

2. Using Little's result, set the overall throughput as

$$\lambda = \frac{m}{\sum_{k=1}^{K} W_k(m) V_k} \tag{5.49}$$

3. For each of the queues Q_k in the network, $k=1, 2, ..., K$, update $N_k(m)$ as

$$N_k(m) = V_k \lambda W_k(m) \tag{5.50}$$

4. For $k=1, 2, ..., K$, calculate the probabilities $p_k(j,m)$ as

$$\begin{aligned} p_k(j,m) &= 1 - \sum_{i=1}^{K} p_k(i,m) && \text{for } j = 0 \\ &= \frac{\lambda p_k(j-1, m-1)}{\mu_k} && \text{for } j = 1, \ldots\ldots, M \end{aligned} \tag{5.51}$$

On conclusion, i.e. after the step for $m=M$ has been done, the results of the above algorithm directly provide the following.

(a) $W_k=W_k(M)$ as the total average time spent by a job in queue Q_k of the network, $k=1, 2, ..., K$. From this, the queueing delay W_{qk} at the queue Q_k may be found.

(b) $N_k=N_k(M)$ as the average total number of jobs in Q_k (including the one in service), $k=1,2, ..., K$. From this, or using Little's result on W_{qk}, we can also find the mean number of jobs N_{qk} waiting in queue at Q_k.

(c) The average system throughput λ will be directly available from Eq. (5.43) at the final step when it is calculated for $m=M$. The average throughput λ_k of the queue Q_k in the network, $k=1, 2, ..., K$, may be found using $\lambda_k=\lambda V_k$.

5.7.3 MVA Algorithm for Closed Network with Multiple Traffic Classes

Unlike the single class case where there is a main loop which increments the number of customers in each iteration cycle, the *Multiple Class* case will have a loop for each class. The basic methodology remains substantially similar to what is done for a single class of customers. We calculate first the average time W_{ik} spent at node i for class k from the average queue sizes obtained in the earlier iteration for one less class k customer. (The equations are given later.) We then determine the throughput (i.e. the internal arrival rate at node i) for each class k using Little's result for the equivalent network as seen by a class k customer entering node i. We finally update the queue length at each node by adding together the queue lengths of each class as obtained in the previous step. These steps have to be executed for all combinations of customer populations. Note that in this case, the population will be denoted by the vector $\boldsymbol{c}$ given by $\boldsymbol{c}=\{c_1,\ldots\ldots, c_R\}$ where R is the number of classes and c_i is the number of customers of class i. We use the following notation.

$W_{ik}(\boldsymbol{c})$= Average time spent by a class k customer at node i when the population vector is $\boldsymbol{c}$

V_{ik} = Visit ratio for class k at node i

R = Number of customer classes

C_i = Total number of customers of class i in the network for $i=1,\ldots\ldots,R$

n = number of queues (nodes) in the network

m_i = number of servers at node i for multiple-server FCFS nodes for $i=1,\ldots,n$

$\boldsymbol{e}_k$ = R-dimensional unit vector

μ_{ik}^{-1} = mean service time (exponential random variable) at node i for class k

The corresponding MVA algorithm is given below.

Step 1. Initialise the following variables to start the recursion process -

$$N_i(0) = 0; P_i\{0,\mathbf{0}\} = 1; P_i\{j,\mathbf{0}\} = 0 \qquad i=1,\ldots\ldots,n \text{ and } j=1,\ldots\ldots, m_i-1$$

Step2. Repeat *Steps 3-6* for $c_1=1,\ldots\ldots,C_1$; $c_2=1,\ldots\ldots,C_2$; ....; $c_R=1,\ldots\ldots,C_R$.
Note that these are the actual steps of the recursion of the MVA

Step 3. *(Mean Value Theorem on each node for each class)*
For $i=1,\ldots,n$; $k=1,\ldots,R$ and if node i is in the route of class k, then

$$W_{ik}(\mathbf{c}) = \frac{1 + N_i(\mathbf{c} - \mathbf{e}_k)}{\mu_{ik}} \qquad \textit{Single Server FCFS, LCFS, PS}$$

$$= \frac{1}{\mu_{ik}} \qquad \textit{IS} \qquad (5.52)$$

$$= \frac{1 + N_i(\mathbf{c} - \mathbf{e}_k) + S_{ik}}{m_i \mu_{ik}} \qquad \textit{Multi Server FCFS}$$

where

$$S_{ik} = \sum_{j=1}^{m_i - 1} (m_i - j) P_i\{j - 1, \mathbf{c} - \mathbf{e}_k\} \qquad (5.53)$$

Step 4. *(Little's Theorem for each class)*
At Node 1 (selected as the reference node), calculate the following for $k=1,......,R$

$$W_k(\mathbf{c}) = \sum_{i=1}^{n} W_{ik}(\mathbf{c}) V_{ik} I_k(i) \qquad (5.54)$$

and

$$\lambda_{1k}(\mathbf{c}) = \frac{c_k}{W_k(\mathbf{c})} \qquad (5.55)$$

where $I_k(i) = 1$ if node i is in the route of class k
$= 0$ otherwise

Step 5. *(Little's Theorem at each node)*
For $i=1,.....,n$, do the following

$$N_i(\mathbf{c}) = \sum_{k=1}^{R} \lambda_{1k} V_{ik} W_{ik}(\mathbf{c}) \qquad (5.56)$$

Step 6. *(Recurrence equation for probability update)*
At each node i $(i=1,......,n)$, do the following for $j=1,.....,m_i-1$

$$P_i\{0,c\} = 1 - \frac{\sum_{k=1}^{R} \frac{\lambda_k(c)V_{ik}}{\mu_{ik}} + \sum_{j=1}^{m_i-1}(m_i - j)P_i\{j,c\}}{m_i}$$

$$P_i\{j,c\} = \frac{1}{j}\sum_{k=1}^{R} \frac{\lambda_k(c)V_{ik}}{\mu_{ik}} P_i\{j-1, c-e_k\} \tag{5.57}$$

where $I_i(m) = 1$ if class m passes through node i
$= 0$ otherwise

Note that at the end of the recursion, we will get the mean time in queue, number in queue and throughput at the queue (with reference to node 1 which has been selected as the reference queue) for each class at each of the nodes of the network.

5.8 Analysis of a Sample Closed Network Using Convolution and MVA Algorithms

We consider here the application of both the convolution algorithm and the MVA algorithm to solve the sample queueing network of Figure 5.12 when there are four jobs circulating in the network, i.e. $M=4$, $K=3$. All the queues are assumed to be single-server queues with respective service rates $\mu_1=1$, $\mu_2=0.5$ and $\mu_3=0.2$. Note that the service times of the jobs in each of the queues Q_1, Q_2 and Q_3 are then exponentially distributed with means 1, 2 and 5, respectively.

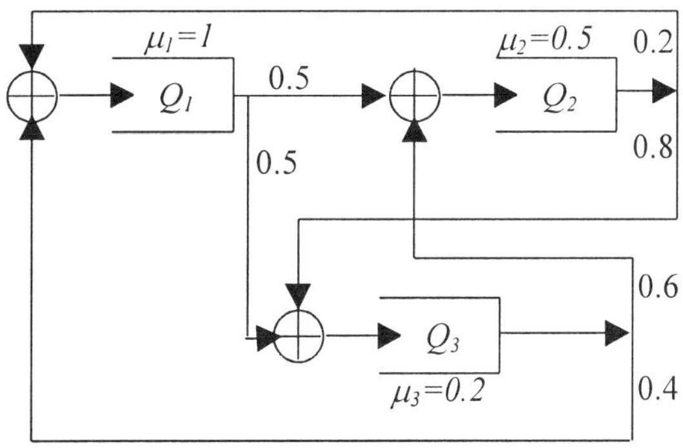

Figure 5.12. A Closed Queueing Network of Single Server Queues, $M=4$

The flow balance equations for this network are

$$\lambda_1^* = 0.2\lambda_2^* + 0.4\lambda_3^*$$
$$\lambda_2^* = 0.5\lambda_1^* + 0.6\lambda_3^*$$

Solving these for a particular solution with $\lambda_1^*=1$, we get

$\lambda_1^*=1,\ \lambda_2^*=1.5385$ *and* $\lambda_3^*=1.7308$ *Relative Throughputs*

The relative utilizations and the visit ratios can also be calculated from the relative throughputs and are found to be

$u_1=1,\ u_2=3.077,\ u_3=8.654$ *Relative Utilizations*

$V_1=1,\ V_2=1.5385,\ V_3=1.7308$ *Visit Ratios*

These may also be computed directly by solving the respective set of linear simultaneous equations.

5.8.1 Using the Convolution Algorithm

In this case, we start the recursion of Eq. (5.35) using the values of the relative utilizations u_1, u_2 and u_3 calculated above with the initial values $g(0,k)=1$ and $g(n,k)=u_1^n$ for $k=1,2,3$ and $n=1,2,3,4$. The following values were obtained.

Table 5.1. Values of g(n,k) for n=0,1,2,3,4 and k=1,2,3

k n	1	2	3
0	1	1	1
1	1	4.077	12.731
2	1	13.545	123.72
3	1	42.678	1113.35
4	1	132.32	9767.26

The value of the normalisation constant for a network with $M=4$ would then be $G(4) = g(4,3) = 9767.26$. Note that the last column of the table actually gives the normalisation constants $G(n)$ for $n=0,1,2,3,4$ in the corresponding rows - these would correspond to a network where there are n circulating jobs.

For $n_1, n_2, n_3 \geq 0$, and $n_1+n_2+n_3=4$, we then get that

$$P(n_1,n_2,n_3)=\frac{1}{9767.26}(3.077)^{n_2}(8.654)^{n_3}$$

as the joint probability of the queue states (i.e. the system state probability) at equilibrium. The actual throughputs of the three queues are found using (5.28) to be

$$\lambda_1=0.114,\quad \lambda_2=0.1754,\quad \lambda_3=0.1973$$

The mean numbers in each queue is computed using Eq. (5.30) as

$$N_1=0.1281,\quad N_2=0.5178,\quad N_3=3.3541$$

5.8.2 Using the MVA Algorithm

In this case we start the recursions of Eqs. (5.42)-(5.44) with the initial values $N_1(0)= N_2(0)= N_3(0)=0$ and do this recursion for $m=1, 2, 3$ and 4. The following values were obtained at each step.

1. For $m=1$

$$W_1(1)=1, W_2(1)=2, W_3(1)=5$$
$$\lambda=0.07855$$
$$N_1(1)=0.07855, N_2(1)=0.2417, N_3(1)=0.6798$$

2. For $m=2$

$$W_1(2)=1.07855, W_2(2)=2.4834, W_3(2)=8.399$$
$$\lambda=0.1029$$
$$N_1(2)=0.11098, N_2(2)=0.3932, N_3(2)=1.49586$$

3. For $m=3$

$$W_1(3)=1.11098, W_2(3)=2.7864, W_3(3)=12.4793$$
$$\lambda=0.1111$$
$$N_1(3)=0.12346, N_2(3)=0.4764, N_3(3)=2.4002$$

4. For $m=4$

$$W_1(4)=1.12346, W_2(4)=2.9528, W_3(4)=17.001$$
$$\lambda=0.114$$
$$N_1(4)=0.12806, N_2(4)=0.51783, N_3(4)=3.35411$$

The average total delay at each queue and the average number in each queue are directly given by the values obtained above for $m=4$. The network throughput $\lambda=0.114$ is also available. Using this and the visit ratios obtained earlier, we can also find the actual throughputs for each of the queues in the network using Eq. (5.46).

5.9 Norton's Theorem for Closed Queueing Networks

Norton's Theorem for the analysis of closed queueing networks draws its inspiration from its analogy with Norton's Theorem in the analysis of electrical circuits where a complex circuit is compactly represented as a current source with parallel impedance driving the load impedance. The current source and its parallel impedance then represents the rest of the circuit other than the load impedance whose effects are required to be studied, i.e. the voltage across the load and the current through it.

Norton's Theorem for closed networks performs an essentially similar function. It is basically a technique to reduce a closed queueing network with K FCFS exponential service queues and M jobs/customers circulating in the network so that the performance of one of the queues (any queue in the network) or the performance of a sub-network of the queues may be easily studied. Following this method, a smaller equivalent network may be obtained by replacing all queues except those in a designated sub-network by a single *Flow Equivalent Server (FES)*. The authors (Chandy, Herzog and Woo) of this method [CHW75] show that for certain types of system parameters, the behaviour of the equivalent network will be exactly the same as that of the original network - this is also known as the Chandy-Herzog-Woo Theorem and is illustrated next. This approach may also be extended as an approximation to a closed network of FCFS queues with general service times. It can also be extended to networks with other service disciplines and may also be used for networks that have several classes of customers. In this section we will, however, limit our discussion to the simple, closed queueing networks with a single class of jobs, which have exponentially distributed service times and probabilistic routing.

In order to illustrate the application of Norton's Theorem to such a closed queueing network consider the example network of Figure 5.13 where we have identified the queue Q_i as the queue whose performance needs to be studied. (Note that we could have also considered a sub-network of queues to be studied instead of just one queue Q_i.)

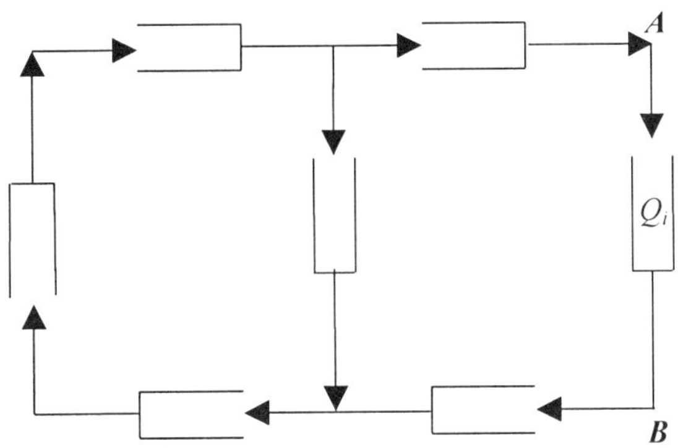

Figure 5.13. Original Queueing Network before Reduction

Consider the situation where we want to characterize the queue Q_i in different ways and see how this affects the performance of the queue and the throughput of the overall network. This, for example, may be done by changing the service rates at Q_i. Obtaining the queueing statistics at Q_i for different characterizations of it will be simplified if the rest of the closed queueing network can be given a more compact representation - preferably as a single queue. The technique of Norton Reduction may be applied to this network to obtain such a compact representation of the rest of the queueing network other than the queue Q_i. The final objective of this objective of this reduction will be to obtain the network given in Figure 5.14 where the sub-network other than Q_i is replaced by a *single queue with a state dependent service rate* $\mu(j)$ where j is the number of jobs in that queue. This queue is also referred to as a *Flow Equivalent Server (FES)* representation of the equivalent queue replacing the rest of the network (other than Q_i). The reason for calling it a flow equivalent server is because it represents that sub-network of queues by a single queue, which is equivalent in terms of the overall flow from that sub-network. It should be noted, however, that the actual nature of the equivalence will depend on the number of jobs that one considers the closed network to have - the equivalent service rate $\mu(j)$ of the flow equivalent server will vary depending on the number of jobs circulating in the network.

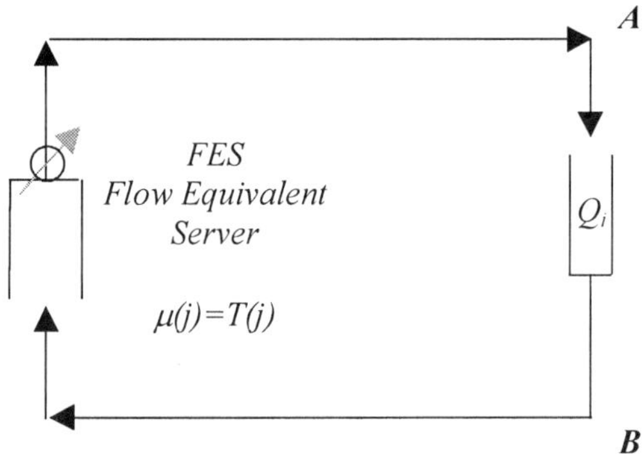

Figure 5.14. Equivalent Network with Flow Equivalent Server

The flow rate $\mu(j)$ of the FES may be calculated as $T(j)$ where this is obtained by shorting A and B and evaluating $T(j)$ as the throughput between these points when there are j jobs circulating in that modified network. This is exactly analogous to way the "short-circuit" current is evaluated in applying Nortons' Theorem to an electronic circuit where the strength of the equivalent current source is taken to be the value of this short-circuit current. This is illustrated in more detail in Figure 5.15. As shown in the figure, the flow rate of the FES is calculated by removing the queue Q_i of interest and connecting its two end points A and B directly, i.e. by short-circuiting the queue. This may also be done by making the service time of Q_i to be zero. The flow between A and B will then be the network's throughput and is calculated as $T(j)$ when there are j jobs circulating in this modified network.

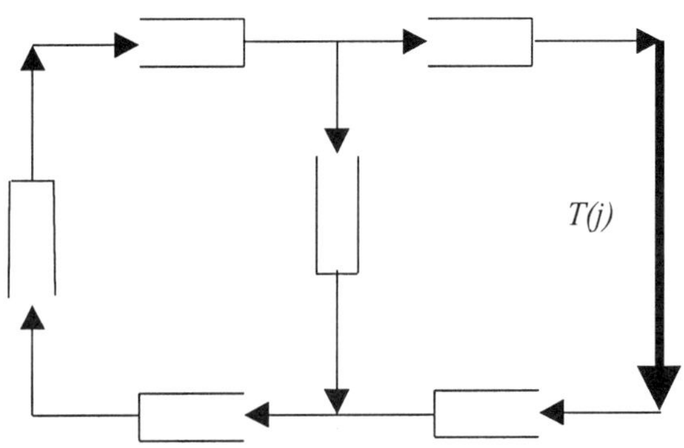

Figure 5.15. Network to Obtain the Flow Rate $T(j)$ of the FES with j Jobs in the Network

In general, the portion of the network that can be isolated is not limited to a single queue and can be a network of queues. Following usual terminology, such a network of queue is called the *designated network* and the remaining network is called the *aggregate*. The aggregate network is reduced to a single *FEC* and the performance of the designated network is studied for variations in the parameters of the designated network. This hierarchical decomposition produces exact results for a large class of networks - i.e. the ones typical referred to as *BCMP Networks* [BCM75] where local balance conditions hold. The procedure is summarized below.

1. Select the *designated sub-network of queues* from the original network that is to be studied. The remaining network is the *aggregate network* to be reduced to a single FES.

2. Generate a queueing network in which the service times at all the queues in the designated network is set to zero, i.e. this is the process of shorting them. Note that the designated network should be selected such that the throughput through all the shorts in the new network should be identical.

3. Solve the above network using any of the known techniques. Solve this for all possible values of the network population, i.e. $j=1,\ldots, M$. The throughput through the short for different populations correspond to the service rate of the FES with that number of jobs in the queue, i.e. $T(j)$.

4. The service rates of the equivalent FES are now available for different values of the number of jobs j circulating in the network. We now consider the equivalent network with the designated network and the FES where the FES replaces the aggregate network. The results for the designated network in this equivalent network will be the same as those in the original network.

Aggregation produces exact results for queueing networks with a product-form solution. However, this approach may be computationally more expensive than using the usual techniques (MVA or Convolution) if the objective is to solve for only one set of parameters for the designated network.

The real advantage of aggregation may actually lie in solving *non-product form queueing networks*. In this case, the usual strategy is to put the components that do not have the product form in the designated network. A FES is then obtained for the aggregate, which contains only those queues that would have a product form solution. The equivalent network may now be solved using either approximate non-product form solution techniques or by simulation. Since solving an actual non-product form network is computationally very expensive, the reduced number of queues in the equivalent model would reduce the computation time. However, in this case, aggregation is only an approximation. This is because the FES cannot exactly model the behaviour of the aggregate as the information on the location of the customers in the aggregate is discarded. However, in many cases of practical interest, the approximation provided by this approach gives acceptable results.

Problems

1. For the following open network with two single server, infinite-buffer queues, Q_1 and Q_2, solve for the throughputs of each queue.

External Arrival Rate to $Q_1 = \lambda$ External Arrival Rate to $Q_2 = 0$
Service Rate of $Q_1 = \mu_1$ Service Rate of $Q_2 = \mu_2$
Routing Probabilities are $P(1,1) = p$, $P(1,2) = 1\text{-}p$, $P(2,2) = q$

2. For the following open network of single-server, infinite capacity queues, Q_1, Q_2 and Q_3, solve for the throughputs of each queue, the average number in each queue and the total delay in the system.

External Arrival at rate λ only to Q_1
Service Rates of Q_1, Q_2 and Q_3 are respectively, μ_1, μ_2, and μ_3
Routing Probabilities are - $P(1,2) = 1$
$P(2,1) = 0.1$, $P(2,2) = 0.55$, $P(2,3) = 0.3$
$P(3,2) = 1$

3. Consider the open network of single-server, FCFS, exponential service time queues shown in Figure 5.16

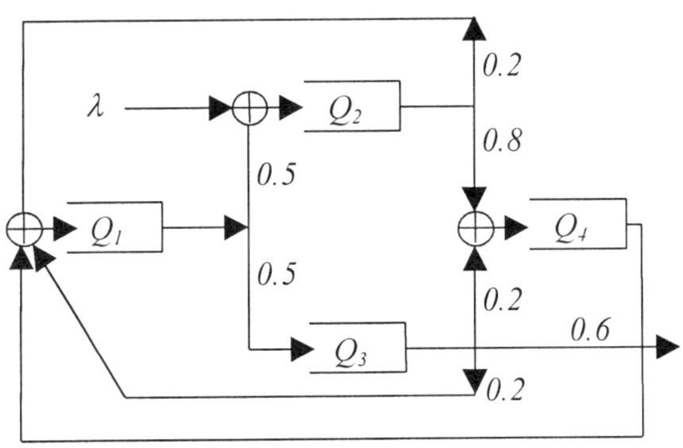

Figure 5.16. Open Queueing Network of Problem 3

The external arrivals are at Q_2 from a Poisson process with average arrival rate λ. The mean service rates are $\mu_1 = \mu_3 = \mu$ and $\mu_2 = \mu_4 = 0.5\mu$.

(a) What will be the maximum value of λ for which the system will be stable?

(b) For $\lambda=0.1$ and $\mu=1$, give the state distribution of the system, the mean number in each queue and the mean time spent in the queueing network by a new customer entering the system.

4. For the following closed network of three single server, infinite-buffer queues, Q_1, Q_2 and Q_3, solve for the visit ratios of each queue -

Routing Probabilities are $P(1,2) = 0.5$ $P(1,3) = 0.5$
$P(2,1) = 1/8$ $P(2,2) = 7/8$
$P(3,1) = 1/4$ $P(3,3) = 3/4$

Assume that μ *is* the service rate of each of the queues in the network, and that the number of jobs circulating in the system is *5*. Solve for the state probability of the overall network. Consider applying the convolution and MVA algorithms to this network as well.

5. Consider the following closed network of single server queues, Q_1, Q_2 and Q_3, with service rates *36*, *3* and *1*, respectively -

Routing Probabilities are $P(1,1) = 1/2$, $P(1,2) = 1/6$, $P(1,3)=1/3$
$P(2,1) = 1$
$P(3,1) = 1$

Assume that there are four customers in the network. Find the relative utilization and actual utilization of each queue, the average number in each queue, the throughput of the overall network and the average delay of each queue using the convolution algorithm and/or the MVA to solve the network.

6. Consider a closed network of four FCFS single-server, infinite-capacity queues, Q_1 - Q_4. The mean service times (exponentially distributed) for Q_1, Q_2, Q_3 and Q_4 are respectively *0.35*, *0.25*, *1.50* and *2.0*. The routing probabilities between the queues are given to be -

From \ To	Q_1	Q_2	Q_3	Q_4
Q_1	0.1	0.5	0.1	0.3
Q_2	0.0	0.0	0.5	0.5
Q_3	0.2	0.2	0.3	0.3
Q_4	0.8	0.0	0.0	0.2

(a) Obtain the visit ratios, V_1-V_4, for each of the queues, taking Q_1 to be the reference queue with $V_1=1$.

(b) Use the MVA Technique to obtain the mean number (waiting and in service) in each queue, and the mean time spent in each queue by an arriving customer when there are a total of *6* customers in the queue.

7. Consider once again the queueing network of Problem 6 with a total of *6* customers in the network, and obtain the following -

(a) Calculate the relative utilizations, u_1-u_4, of each queue, assuming $u_1=1$

(b) Apply the Convolution Algorithm to find the Normalisation Constant.

(c) Calculate the actual utilization of each queue.

8. Consider the closed network of single-server queues shown in Figure 5.17. The mean service times (exponentially distributed) at the queues Q_1, Q_2 and Q_3 are given to be *0.028* seconds, 0.*040* seconds and *0.400* seconds, respectively. Assume that there are 6 customers in the network.

(a) Calculate the normalisation constant and give the expression for the state distribution of the network.

(b) Find the actual throughput of each queue and the throughput of the network.

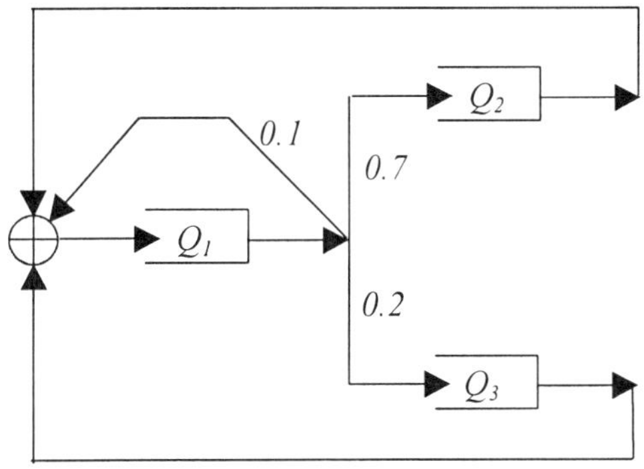

Figure 5.17. Closed Queueing Network for Problem 8

(c) Use the state probability distribution to find the mean number in each queue and the mean time spent by a job in each queue.

(d) Obtain the results of (c) directly by solving the network using the MVA algorithm.

9. Consider the closed queueing network of single server queues with exponentially distributed service times, as shown in Figure 5.18.

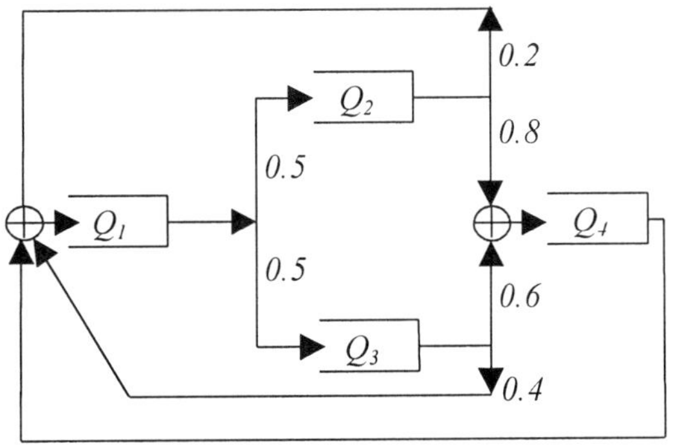

Figure 5.18. Closed Queueing Network of Problem 9

The average service rates of the queues Q_1, Q_2, Q_3 and Q_4 are *1.0, 1.0, 0.5* and *0.5*, respectively. The system has a total user population of 6.

Use the convolution algorithm to obtain the normalisation constant and the state probability distribution of the network. Use these results as well as the MVA algorithm to obtain the actual throughput of each queue, the overall network throughput, the mean number in each queue and the mean of the total time spent in each queue.

Chapter 6

Advanced Queueing Networks

Approximation Techniques and Algorithms for Open and Closed Queueing Networks

Exact analytical methods are available for studying open and closed queueing networks with product-form solutions. Examples of such networks and the methods for obtaining exact analytical results for them were considered in Chapter 5. These results are applicable only if the queues satisfy all the restrictive conditions that are required on their respective arrival and service processes. These conditions may be quite restrictive and may not allow us to use these techniques under more general conditions.

In this chapter, we consider various approximation techniques that may be used to study queueing networks of different kinds, which cannot be handled by the exact analytical methods of Chapter 5. Some of these are based on extensions which assume that even though the product-form solution may not be satisfied in such networks, it still holds as a remarkably good approximation. Other approximate approaches and some extended queueing models (such as mixed networks and fork/join queues) are also considered in this chapter. In general, these have been observed to provide good results. However, the accuracy of the results obtained using these methods cannot really be guaranteed. If these methods are used, the user should check the results obtained for typical cases using either simulations or other methods of approximation.

6.1 Mixed Queueing Networks

Mixed queueing networks are networks with multiple customer classes. The different customer classes are such that the network may be considered open for some classes while it is closed for the others. The two different types of customer classes may however share service at one or more queues in the network. A brief description of such networks was given earlier in Section 5.1.3.

Each of the open classes in the network is characterised by the average arrival rate of that class entering the network from outside - this will also be the average arrival rate of customers of this class leaving the network altogether after service has been obtained. Each closed class is characterised by the number of jobs of that class circulating in the network; this is generally given as part of a population vector (i.e. population of each closed class) given for all the closed classes in the network. In addition to these parameters, the other typical input parameters such as routing probabilities for each class (assuming static routing) and mean service times for each class at each of the network nodes (assuming exponentially distributed service times) must also be given.

Lazowska et al [LZS84] provides an approximate algorithm for analysing a queueing network of this type. This may be applied when the network and the queues are loaded such that equilibrium conditions exist in the network, the service times are assumed to be exponentially distributed and the arrival processes of the open classes are Poisson in nature. In this algorithm, the utilisation for each of the open classes at each node is computed first. These are then used to compute the net utilisation of each node considering only the open classes. The performance parameters of the closed classes are subsequently computed by eliminating the open classes and converting the mixed network model into a *closed model with inflated service demands for the closed classes*. The inflation factor used at node i for this is $(1-U_{i,\{O\}})$, i.e. (1 - utilisation due to all the open classes at node i), which is given by

$$U_{i,\{O\}} = \sum_{c=1}^{O} U_{i,c} \tag{6.1}$$

where $U_{i,c}$ is the utilisation of class c at node i. We should observe that this is the percentage of time the server is not used by the open classes and is therefore the time available for the closed classes to get service in the queue. (This service inflation is a basic approximation inherent in this technique.) The performance of the closed classes is analysed using the standard solution techniques discussed earlier, using the inflated service times for the closed class customers as given above. The performances of the open classes are then computed by considering the actual mean queue length of the closed classes at each node in the network. Note that this method may also be used when either the number of open classes or the number of closed classes is zero as in those cases it reduces to the usual Jackson Network models for closed and open networks, respectively. For convenience, we summarise below the basic assumptions that are needed to apply this method.

- All the queues in the network have infinite buffer capacity.
- The service disciplines for all classes are FCFS at all the nodes in the network.
- There can be creation or combination of jobs for the open classes. (This has not been considered here.)
- All the classes are independent of each other.
- There must be an external arrival for each of the open classes. These processes must be Poisson.
- The service time distributions for all the classes in the network are exponential.
- The routing probabilities for the jobs in the network are static for all the job classes. However, these may be different for the different classes.

The algorithm described subsequently requires the following input parameters.

Input Parameters

R total number of classes in the mixed model
O total number of open classes in the mixed network
C total number of closed classes in the mixed network
(Note that $R = O + C$)
N number of nodes in the mixed model
m_i number of servers at node i in the mixed network for $i = 1,2,\ldots, N$
$\mathbf{C}$ population vector of the closed classes $= \{ C_1, C_2, \ldots, C_C \}$
$[\boldsymbol{Q_k}]_{NxN}$ routing probability matrix for $k=1, 2, \ldots, R$ with $q_{k,(i,j)}$ as the routing probabilities for $i,j=1, 2,\ldots, N$

For the open classes:

$\lambda_{0i,c}$ mean external arrival rate of class c at node i for $c=1,2,\ldots,O$ and $i=1,2,\ldots, N$
τ_{ic} mean service time distribution of class c at node i, for $c=1,2,\ldots,O$ and $i=1,2,\ldots, N$
γ_{ic} multiplication factor of class c at node i for $i=1,2,\ldots, N$ and $c=1,2,\ldots,O$

For the closed classes:

τ_{ic} mean service time distribution of class c at node i , for c=1,2,....,C and i=1,2,..., N

The algorithm provides the following output parameters regarding the queues in the queueing network.

Output Parameters

$N_{i,\ open}$ mean number of open class jobs at node i
$N_{i,\ closed}$ mean number of closed class jobs at node i

For the open classes:

λ_{ic} net arrival of class c at the node i for i=1,2,......, N and c=1,2,......,O

$N_{i,\ open,\ c}$ mean queue length of class c at the node i for i=1,2,......, N and c=1,2,......,O
$W_{i,\ open,\ c}$ mean delay of class c at the node i for i=1,2,......, N and c=1,2,......,O
$dep_{i,\ open,\ c}$ departure rate of class c at the node i for i=1,2,......, N and c=1,2,......,O
$V_{i,\ open,\ c}$ visit counts of class c at the node i for i=1,2,......, N and c=1,2,......,O
$U_{i,\ c}$ utilization of class c at the node i for i=1,2,......, N and c=1,2,......,O

For the closed classes:

$N_{i,\ closed,\ c}$ mean queue length of class c at node i for i=1,2,..., N and c=1,2,...,C
$W_{i,\ closed,\ c}$ mean delay of class c at node i for i=1,2,..., N and c=1,2,...,C
$\lambda_{i,\ closed,\ c}$ throughput of class c at node i for i=1,2,..., N and c=1,2,......,C
$V_{i,\ closed,\ c}$ visit ratios of class c at node i for i=1,2,..., N and c=1,2,...,C

The actual algorithm, described earlier, is given below.

Algorithm

Step 1. Solve the following equations, to get the net arrivals to the nodes for each of the open classes

$$\lambda_{ik} = \lambda_{Oi,k} + \sum_{j=1}^{N} \lambda_{ik} q_{k,(j,i)} \qquad i=1,2,\ldots, N \text{ and } k=1,2,\ldots,O \tag{6.2}$$

$$U_{i,k} = \frac{\lambda_{ik}\tau_{ik}}{m_i} \qquad i=1,2,\ldots, N \text{ and } k=1,2,\ldots,O \tag{6.3}$$

$$U_{i,\{O\}} = \sum_{k=1}^{O} U_{i,k} \qquad i=1,2,\ldots, N \text{ and } k=1,2,\ldots,O \tag{6.4}$$

$$dep_{i,open,k} = \lambda_{ik}\gamma_{ik}\left(1 - \sum_{j=1}^{N} q_{k,(i,j)}\right) \qquad i=1,2,\ldots,N \text{ and } k=1,2,\ldots,O \tag{6.5}$$

$$throughput\ (k) = \sum_{i=1}^{N} \lambda_{ik} \qquad k=1,2,\ldots,O \tag{6.6}$$

$$V_{i,open,k} = \frac{\lambda_{ik}}{throughput\ (k)} \qquad i=1,2,\ldots,N \text{ and } k=1,2,\ldots,O \tag{6.7}$$

Step 2. Inflate the service times of the closed classes by a factor of $[1\text{-}U_{i,\{O\}}]$ at each node i.

$$\tau^{*}_{ik} = \frac{\tau_{ik}}{1 - U_{i,\{O\}}} \qquad i=1,2,\ldots,N \text{ and } k=1,2,\ldots,C \tag{6.8}$$

We should then solve the closed network as a multiple-class, closed network using the MVA method of Section 5.7.

Step 3. The mean delays and the queue lengths for each of the O open classes are then computed at each of the N nodes of the network.

$$W_{i,open,k} = \frac{\tau_{ik} V_{i,open,k}(1 + Qlength_{i,\{C\}})}{1 - U_{i,\{O\}}} \quad i{=}1,2,\ldots,N \text{ and } k{=}1,2,\ldots,O \quad (6.9)$$

where,

$$\begin{aligned} Qlength_{i,\{C\}}) &= N_{i,closed} \\ &= \sum_{k=1}^{C} N_{i,closed,k} \end{aligned} \qquad i{=}1,2,\ldots,N \qquad (6.10)$$

and

$$N_{i,open,k} = (throughput(k))W_{i,open,k} \qquad i{=}1,2,\ldots,N \qquad (6.11)$$

$$N_{i,open} = \sum_{k=1}^{O} N_{i,open,k} \qquad i{=}1,2,\ldots,N \qquad (6.12)$$

6.2 The GI/G/m Approximation for the Approximate Analysis of Open Queueing Networks (the QNA technique)

The GI/G/m approximation described here is an example of a method using *Parametric Decomposition* where the individual queueing nodes are analysed in isolation based on their respective input and output processes. This method has been used by Whitt in [Whi83] for the *Queueing Network Analysis* (QNA) software.

In this model, the arrival process to each queue is assumed to be a generalised inter-arrival (GI) process. The service times may have any general distribution. The approximation made by this approach is that only the *mean* and the *squared coefficient of variance (SQV= variance/(mean)²)* of the inter-arrival times and service times are required for our calculations - this is the reason why this kind of method is sometimes referred to as a *two-moment method.* The queueing network involves nodes where customer streams may join (i.e. the individual flows get combine before entering the queue/node) or customers leaving a node may be split into different streams in a probabilistic fashion. We have encountered these kinds of splitting and superposition earlier in the context of our discussions on basic queueing networks. An example of superposition and splitting is shown in Figure 6.1.

It is also possible to use this approach to handle networks where there is feedback. However, *immediate feedback*, where a fraction of the output of a particular queue enters the queue once again, needs special treatment because of the close correlation this will cause between the input and output of the queue. This has been described subsequently. The method also allows the inclusion of a *multiplication factor* v, $v > 0$, by which a job finishing service at a queue becomes v jobs before the subsequent routing is done. Note that if $v < 1$, then this really corresponds to attenuation of the number of jobs departing from a queue. The method was first proposed by Whitt for use in his *Queueing Network Analysis* (QNA) software [Whi83] and has been shown to provide remarkably good results for general, open queueing networks.

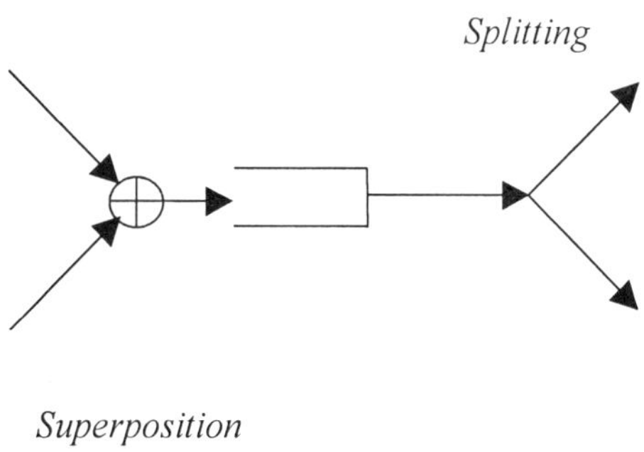

Figure 6.1. Superposition and Splitting in a Queueing Network

In order to apply this method, we assume that the arrival process to a network node is *renewal* in nature, in which the arrival intervals are all, independent, identically distributed (i.i.d.) random variables. It may also be possible to approximate a process that is actually non-renewal in nature by a suitable renewal process with the same first and second moments. Before the actual detailed analysis of the queueing network is done, the method first removes immediate feedback in a queue by suitably modifying its service time and the routing probabilities of the jobs leaving the queue. This is done for all the queues that have immediate feedback in the network and the modified network is used in the subsequent analysis. The algorithm then calculates the mean and variance of the internal arrival processes to the individual nodes/queues by solving a set of linear equations for the mean flows. This can be done as the queueing network is assumed to be in equilibrium and hence the mean flow entering a queue must equal the mean flow leaving the queue. These flow balance equations depend on the whether the flow is split, superposed or both split and superposed at the nodes. For each queue in the network, it then replaces this internal arrival process by a renewal process with equivalent first and second moments. This effectively decomposes the queueing network into individual subsystems (i.e. the queues or nodes) that can be examined separately. Each of these would be a standard GI/G/m queue characterised by the first two moments of its arrival process and service time. Approximate two-moment based formulas are available for calculating the congestion measures in such a queue. These would be the delay and the queue length at each queue. Note that if a queue has been modified by the removal of its immediate feedback, then the delay and queue length parameters of the queue have to be modified once again to get the results for the queue when immediate feedback is present. Once the individual queue congestion measures are known, network performance measures may be suitably computed. We summarise below the basic assumptions of this method.

- The queueing network is an *open* one
- The individual queues may have one or more servers and has infinite waiting space so that there is no blocking or loss anywhere in the system
- The service discipline is FCFS in nature
- The approach is described here for a single class of customers with probabilistic routing. (Whitt's original paper also discusses using this for multi-class systems where the routing is deterministic in nature.)
- The routing matrix is static and does not change over time
- Whitt's original paper also incorporates the concept of a *multiplication factor* that can be applied at a node. For example, a job sent by a node may get split into several jobs to provide this multiplication effect. From that point onwards these jobs may travel independently in the network.

The following input parameters are required for this approach.

Input Parameters

K	number of nodes/queues in the network
m_i	number of servers at node i
λ_{ji}	flow leaving Q_j which goes to Q_i
λ_{0i}	external arrival rate to node i

c_{oi}^2 squared coefficient of variation (SQV) of the inter-arrival time of the external arrival process to node i

Note that $SQV = \dfrac{\text{var}\,iance}{(mean)^2}$

τ_i average service time at node $i = \mu_i^{-1}$

c_{si}^2 squared coefficient of variation (SQV) of the service time at node i

$P=[p_{ij}]$ Probability transition matrix where p_{ij} is the probability that a job finishing service at node i goes to node j

v_i Multiplication factor at the output of node i

6.2.1 Network Reconfiguration by Immediate Feedback Removal

Queues in the network which have *immediate feedback* ($p_{ii}>0$) as shown in Fig. 6.2 pose a problem in this method which needs to be handled separately. This is because the GI/G/m algorithm basically assumes that the input and output of the queue are uncorrelated - this would not be true if there is immediate feedback around the queue. An example of this has been shown in Figure 6.2. Here, a fraction p_{ii} of the output traffic from Q_i is fed back to this queue itself so that the net arrival process to the queue is the sum of the external arrivals from Λ and the fed back portion, i.e. $p_{ii}\lambda_i$.

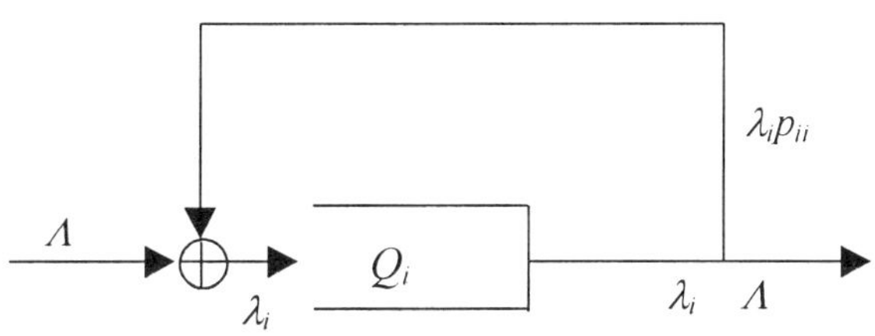

Figure 6.2. Queue with Immediate Feedback

The approach followed to eliminate this immediate feedback at the queue is to suitably adjust the service times at the queue and the transition probabilities for jobs leaving the queue. This is done to take into account the fact that some of the jobs get routed back to this queue because of the presence of the immediate feedback. This would leave a queue without immediate feedback as shown in Figure 6.3.

Assume that the original service parameters for Q_i are

Mean Service Time = $\tau_{i,U}$
SQV of Service Time = $c_{si,U}^{\;2}$

and the original routing probabilities are $p_{ij,U}$ $j=1,.....,K$ with $p_{ii,U}>0$.

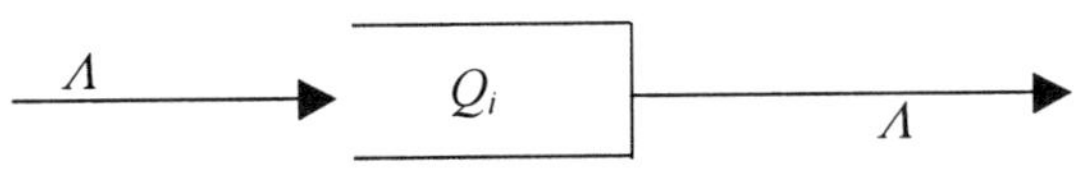

Figure 6.3. Queue after Removal of Immediate Feedback

Removing this immediate feedback from Q_i, we will get the modified service parameters as

$$\tau_{i,M} = \frac{\tau_{i,U}}{1 - p_{ii,U}} \tag{6.13}$$

$$c_{si,M}^2 = p_{ii,U} + (1 - p_{ii,U})c_{si,U}^2 \tag{6.14}$$

with the new routing probabilities $p_{ij,M}$ for jobs leaving Q_i given by

$$
\begin{aligned}
p_{ij,M} &= \frac{p_{ij,U}}{1-p_{ii,U}} \qquad j \neq i \\
&= 0 \qquad\qquad\quad j = i
\end{aligned}
\tag{6.15}
$$

Effectively, this reconfiguration combines the multiple services that a job may receive at Q_i (because of the feedback with probability $p_{ii}=p_{ii,U}$) into a single large service interval. This reconfigured queue without immediate feedback is used subsequently for solving the queueing network. The performance measures obtained by this analysis for *reconfigured queue* would have to be modified once again to get the results for the actual queue, i.e. the one prior to the reconfiguration removing immediate feedback. Let $W_{qi,U}$ be the mean waiting time for a job in the original queue Q_i (unmodified, prior to immediate feedback removal) and let $W_{qi,M}$ be the mean waiting time for a job in the modified Q_i (after immediate feedback removal, i.e. the queue in the queueing network which is solved, as described subsequently, after immediate feedback removal). After the solution process (in the network without immediate feedback) yields $W_{qi,M}$ as the queueing delay for Q_i, we can then obtain the actual queueing delay in the original unmodified Q_i as

$$
W_{qi,U} = (1 - p_{ii,U})W_{qi,M} \tag{6.16}
$$

The SQV of this waiting time may also be computed as shown in Whitt's paper [Whi83]. The average rate of the actual arrival process $\lambda_{i,U}$ (i.e. $\lambda_{i,U}=\lambda_i$) entering Q_i (actual queue before reconfiguration done for immediate feedback removal) may also be calculated from the value $\lambda_{i,M}$ obtained by solving the queueing network after immediate feedback removal. This will be given by

$$
\lambda_{i,U} = \frac{\lambda_{i,M}}{(1 - p_{ii,U})} \tag{6.18}
$$

We have mentioned earlier, that the procedure for immediate feedback removal at a queue, essentially amounts to combining the multiple services that a job will get at the queue (because of the feedback) into one longer service interval. Note that if the queue was FCFS in nature, then we may have a situation where the job which is fed back, rejoins the queue behind all the other customers waiting for service. The model used here for immediate feedback removal does not capture this detail. However, this is not expected to affect the mean performance parameters of the queue.

With immediate feedback removed from the queues of the network, from where ever it may be present, the QNA approximation then considers networks where none of the queues have any immediate feedback. In the following, we consider a queueing network of this kind, assuming that immediate feedbacks around any of the original queues have already been appropriately removed and the network has been suitably reconfigured. It can be easily seen that in such a network, the traffic flows (i.e. the arrival processes to the various queues and arrivals/departures from/to outside the network) require one of the following three operations. In each case, the resulting process will be described as a *General Inter-arrival (GI)* process and will be described using a two-moment approach based on the mean and SQV of its inter-arrival times. (The mean flow rate will be (mean service time)$^{-1}$.) These flow operations are listed below.

1. Superposition of GI streams approximated as a GI stream
2. Splitting a GI stream probabilistically to obtain several GI streams
3. GI stream passing through a Queue with its output process approximated as another GI stream

We consider these in detail next for calculating the mean flow rate and the SQV (of the inter-arrival times) of the internal flows.

6.2.2 Calculating the Parameters (Mean and SQV) of the Internal Flows

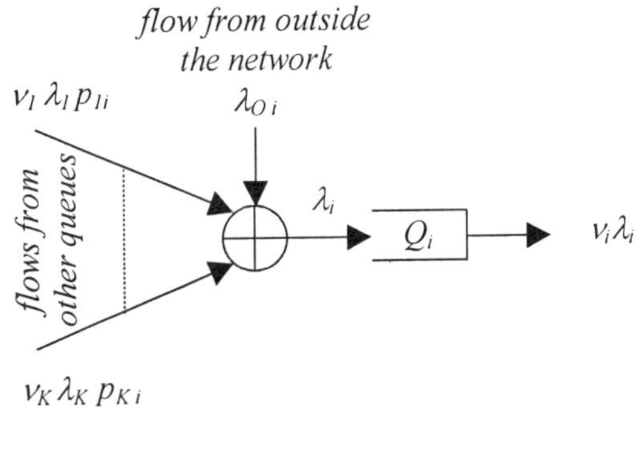

Figure 6.4. Internal Flow Parameter Calculations

In Figure 6.4, we have shown a typical queue Q_i in a network of the type being considered here. Note that λ_i is the internal flow to Q_i which comes both from the flows $\lambda_1, \ldots, \lambda_K$ from the other $(K-1)$ queues in the network (after appropriate multiplication and routing) as well as the flow λ_{0i} entering the network from outside at Q_i. It should be noted that following Whitt's original formulation, we have allowed the use of a multiplication factor v_k for the flow leaving Q_k. This implies that if the flow rate entering Q_k is λ_k, then the flow rate leaving this queue would be $v_k\lambda_k$. Incorporating this multiplication factor, the flow balance equations may then be written as

$$\lambda_i = \lambda_{0i} + \sum_{j=1}^{K} \lambda_j v_j p_{ji} \qquad \text{for } i=1,\ldots\ldots, K \tag{6.19}$$

Note that p_{ii}, $i=1,\ldots\ldots, K$ in the above equation will actually be zero since immediate feedback has already been removed. However, this term has still been included for ease of notation. The mean flow rate entering each of the K queues in the network may then be obtained by solving Eq. (6.19), given the external flows, the multiplication factors (if any) at the output of each queue and the routing probabilities. Note also that if m_i is the number of servers at Q_i then we can write

$$\text{Server Utilization at } Q_i = \rho_i = \frac{\lambda_i \tau_i}{m_i} \tag{6.20}$$

For calculating the *internal arrival SQV*, c_{ai}^2 at Q_i, we again get a system of equations but the derivation of these equations is not as straight forward as for the flows. This is because, the superposition of the K arrival processes (which we assumed to be renewal in nature) does not yield a renewal process except when each of the component processes are individually Poisson in nature. This means that we cannot get the nth moment of the superposition stream by simply adding the nth moments of the individual streams. Instead, approximations will have to be made to derive the variance and other moments of the inter-arrival times.

Consider the internal flow process into Q_i. We want to replace this by a renewal process with the same first and second moments. The basic approximation process here is to use the nth partial sum S_n (which is the sum of the first n arrival intervals) of the superposition process and use its first two moments to approximately define the approximating renewal process.

Let $M_j(X)$ denote the jth moment of some random variable X. If H is the random variable which represents the inter-arrival time of the approximating renewal process, then the moments of H are calculated from those of the nth partial sum to be

$$M_j(H) = \frac{M_j(S_n)}{n}$$

We should then first calculate the moments of S_n and then divide that by n to get the approximate moments of H. For doing this, the *Stationary Interval Method* uses $n=1$ which does not take interdependence between subsequent intervals into account but is a simple and workable choice in many cases. The *Asymptotic Method*, on the other hand uses $n=\infty$ considering the asymptotic behaviour of the process as time tends to infinity. In general, any one method does not work well for the entire range of traffic intensities. Intuitively, it may be observed that the *stationary interval method* works well at low traffic intensities as here the inter-arrival times would be large enough to reduce the approximation errors. Under heavy traffic conditions, the *asymptotic method* is found to be more accurate. Therefore a hybrid combination of the moments obtained using both the methods probably offers the best consensus choice. For our applications, using this kind of a hybrid poses one problem. The stationary interval method of approximating a superposition stream by a renewal process gives a variance which a non-linear function of the variances of the component processes. This would lead to the unfortunate situation where a non-linear set of equations would need to be solved to get the second moment of the superposition process (i.e. the internal arrival process). Fortunately, a convex combination of the variance obtained the asymptotic method and that of an exponential (which would have a variance coefficient of unity) is found to perform well and has been suggested for use by Whitt in [Whi83].

For this approximation, consider the departure process from Q_j. Approximating this point process by a renewal process, the stationary interval method would give the following values for the SQV $(c_{dj}^2)_S$. (Note that the result given for the GI/G/m queue is really an approximation.)

$$\left(c_{dj}^2\right)_S = \rho_j^2 c_{sj}^2 + (1-\rho_j^2)c_{aj}^2 \qquad \text{for a GI/G/1 queue}$$

$$\left(c_{dj}^2\right)_S = 1 + (1-\rho_j^2)(c_{aj}^2 - 1) + \rho_j^2 m_j^{-0.5}(\max\{c_{sj}^2, 0.2\} - 1) \qquad \text{for GI/G/m queue}$$

Here, the subscript S denotes the fact that this is calculated using the stationary interval method.

The number of departures d_t in time interval $(0, t)$ from the queue would be given by $d_t=a_t - n_t$. Here a_t is the number of arrivals in that interval and n_t is the number in the system at time t. With t tending towards ∞ (as for the asymptotic approximation), n_t would tend towards a steady state value if the queue is stable. This would mean that d_t would follow a_t and therefore the asymptotic approximation for c_{dj}^2 would be c_{aj}^2 itself, i.e.

$$\left(c_{dj}^2\right)_A = c_{aj}^2$$

Here, the subscript A denotes the fact that this is calculated using the asymptotic method.

At the output of Q_j, the flow gets split probabilistically according to the probabilities p_{ji}, $i=1,....,j$ $i\neq j$. If a renewal process with SQV c^2 is split into k streams according to probability p_i, $i=1,....,k$, then the SQV of the ith stream will be

$$c_i^2 = p_i c^2 + (1 - p_i) \tag{6.21}$$

Using this and both the stationary interval and asymptotic methods, we get

$$\left(c_{ji}^2\right)_S = p_{ji}\left(c_{dj}^2\right)_S + (1 - p_{ji}) \tag{6.22}$$

$$\left(c_{ji}^2\right)_A = p_{ji}\left(c_{dj}^2\right)_A + (1 - p_{ji}) \tag{6.23}$$

Therefore the SQV of the stream from j to i, computed as a hybrid of the above two methods is

$$c_{ji}^2 = \alpha_{ji}\left(c_{ji}^2\right)_A + (1 - \alpha_{ji})\left(c_{ji}^2\right)_S \tag{6.24}$$

The asymptotic SQV can then be found as

$$\left(c_{ai}^2\right)_A = \sum_{j=0}^{K} \frac{\lambda_{ji}}{\left[\sum_{k=0}^{K} \lambda_{jk}\right]} c_{ji}^2 = \sum_{j=0}^{K} q_{ji} c_{ji}^2 \tag{6.25}$$

with

$$q_{ji} = \frac{\lambda_{ji}}{\left[\sum_{k=0}^{K} \lambda_{jk}\right]} \tag{6.26}$$

Note that q_{ji} represents the proportion of arrivals to Q_i that come from Q_j. Using the above and the fact that a convex combination of the asymptotic value and an exponential SQV (=*1*) gives a good approximation to the actual SQV, we get

$$c_{ai}^2 = \varpi_i \left(c_{ai}^2\right)_A + (1 - \varpi_i) \tag{6.27}$$

This may also be written in the following convenient form

$$c_{ai}^2 = u_i + \sum_{j=1}^{K} c_{aj}^2 v_{ji} \tag{6.28}$$

where

$$u_i = 1 + w_i \left[(q_{0i} c_{0i}^2 - 1) + \sum_{j=1}^{K} q_{ji} \left((1 - p_{ji}) + v_j p_{ji} \rho_j^2 x_j \right) \right]$$

$$v_{ji} = w_i q_{ji} p_{ji} v_j (1 - \rho_j^2)$$

$$x_j = 1 + \frac{\max(c_{sj}^2, 0.2) - 1}{\sqrt{m_j}}$$

$$w_i = \frac{1}{1 + 4(1 - \rho_i)^2 (\gamma_i - 1)}$$

$$\gamma_i = \frac{1}{\sum_{i=0}^{K} q_{ij}^2}$$

This set of linear equations for $i=1,.....,K$ may then be solved to get c_{ai}^2.

Once the parameters of the internal flows have been found (using the above equations), we can then solve for the congestion measures (i.e. the queueing parameters) at each queue assuming the queues to be independent of each other. The output parameters that may be computed are described next.

6.2.3 Output Parameter Calculations for the Node Q_i

The following parameters of interest may be calculated for each node in the network. We consider here the results for the ith queue Q_i.

Mean Waiting Time W_{qi} and Related Parameters N_{qi}, W_i, N_i

If Q_i is a GI/G/1 queue, the following approximation (Kramer and Langenbach-Belz approximation) is used to get the mean waiting time in queue.

$$W_{qi} = \frac{\tau_i \rho_i (c_{ai}^2 + c_{si}^2)\beta}{2(1-\rho_i)} \qquad \text{(for GI/G/1 queue)} \qquad (6.29)$$

with

$$\begin{aligned} \beta &= \exp\left(-\frac{2(1-\rho_i)(1-c_{ai}^2)^2}{3\rho_i (c_{ai}^2 + c_{si}^2)} \right) & c_{ai}^2 < 1 \\ &= 1 & c_{ai}^2 \geq 1 \end{aligned} \qquad (6.30)$$

For the SQV of the waiting time in the ith queue, the approximation recommended is

$$c_{Wqi}^2 = \frac{c_{Di}^2 + 1 - \chi_i}{\chi_i} \qquad (6.31)$$

with

$$\chi_i = P\{W_{qi} > 0\} = \rho_i + (c_{ai}^2 - 1)\rho_i (1-\rho_i) h(\rho_i, c_{ai}^2, c_{si}^2)$$

$$\begin{aligned} h(\rho_i, c_{ai}^2, c_{si}^2) &= \frac{1 + c_{ai}^2 + \rho_i c_{si}^2}{1 + \rho_i (c_{si}^2 - 1) + \rho_i^2 (4c_{ai}^2 + c_{si}^2)} & c_{ai}^2 \leq 1 \\ &= \frac{4\rho_i}{c_{ai}^2 + \rho_i^2 (4c_{ai}^2 + c_{si}^2)} & c_{ai}^2 \geq 1 \end{aligned}$$

$$c_{Di}^2 = 2\rho_i - 1 + \frac{4(1-\rho_i)\xi_{si}}{3(c_{si}^2+1)^2}$$

$$\xi_{si} = 3c_{si}^2(1+c_{si}^2) \qquad c_{si}^2 \geq 1$$

$$= (2c_{si}^2+1)(c_{si}^2+1) \qquad c_{si}^2 < 1$$

If Q_i is a GI/G/m queue, an approximation based on heavy traffic limit theorems is suggested. This recommends modifying the corresponding results from the M/M/m queue to get approximate values of the mean and SWV of the queue Q_i as

$$W_{qi} = \left(\frac{c_{ai}^2 + c_{si}^2}{2}\right) W_{qi}^{M/M/m} \quad \text{(for GI/G/m queue)} \tag{6.32}$$

Here $W_{qi}^{M/M/m}$ is the waiting time in queue for the corresponding M/M/m queue. We approximate the SQV of the waiting time to be the same as that of an M/M/m queue with the same set of input parameters, i.e.

$$c_{Wqi}^2 = c_{Wqi}^{2\;M/M/m} \quad \text{(for GI/G/m queue)} \tag{6.33}$$

Once W_{qi} has been found, the other related parameters for Q_i may be found in the usual way as follows

Mean total time spent in Q_i by a job (on every arrival) $=W_i=W_{qi}+\tau_i$

Mean number of jobs waiting in Q_i, prior to service $= N_{qi}=\lambda_i W_{qi}$

Mean number of total jobs in Q_i (waiting and in service) $= N_i=\lambda_i W_i$

Visit Ratio for Q_i

The visit ratio V_i for Q_i is defined as usual to be the average number of visits to node i by a job during the entire time it spends in the network. This will be given by -

$$V_i = \frac{\lambda_i}{\sum_{j=1}^{K} \lambda_{0j}} \tag{6.34}$$

Sojourn Time at Q_i

The Sojourn Time T_i is the average total time an arbitrary job spends in node *i*, until it departs from the system. This will be

$$T_i = V_i(\tau_i + W_{qi}) \tag{6.35}$$

6.2.4 Network Sojourn Time and Departure Rate

The Sojourn Time *T* for the whole network will be the mean total time that a job entering the network spends inside the network before its ultimate departure. This will be -

$$T = \sum_{j=1}^{K} T_j \tag{6.36}$$

If the multiplication factors of all the nodes are equal to 1, then the total output flow rate from the network will equal the net input flow rate. In this case, the total departure rate *d* from the network will be given by

$$d = \sum_{i=1}^{K} \lambda_{0i} \qquad v_i = 1 \tag{6.37}$$

For the general case of $v_i \neq 1$, we get that

$$d = \sum_{i=1}^{K} \lambda_i v_i (1 - \sum_{j=1}^{K} p_{ij}) \qquad v_i \neq 1 \tag{6.38}$$

6.3 Fork/Join Queues in Open and Closed Networks of Infinite Capacity Queues

In a fork/join node (or queue), an entering job [KiA89], lLiP91], [NeT88] is decomposed to be serviced in parallel by a number of sibling queues. This is referred to as the *forking process* where one job is offered to the two or more sibling queues for every job which actually enters the node. Once the jobs get their desired service at all the sibling queues, they are recombined into one job once again by the *joining process* before departing from the fork/join node. Note that for every job entering the fork/join node, only one job leaves the node (after all the sub-jobs at all the sibling queues get completed), even though multiple copies of the job (one for each sibling queue) will get created inside the fork/join node. Examples of such fork/join nodes with k sibling queues are shown in Figures 6.5 and 6.6.

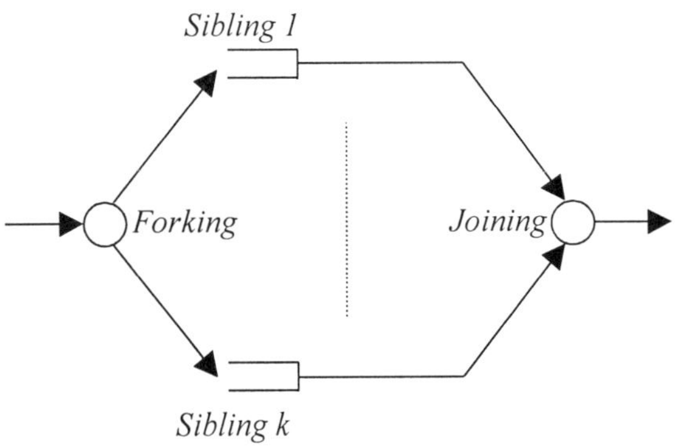

Figure 6.5. Fork/Join Node without Synchronising Queues

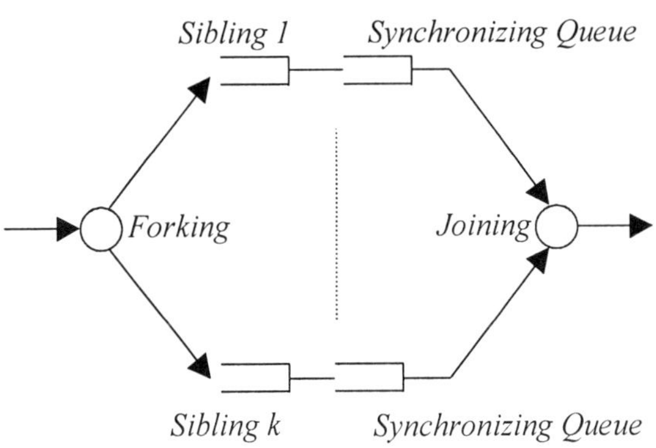

Figure 6.6. Fork/Join Node with Synchronising Queues

As shown for the two cases illustrated in Figs. 6.5 and 6.6, two kinds of fork/join nodes may be considered. These are *fork/join nodes without synchronising queues* and *fork/join nodes with synchronising queues*. These two kinds of queues operate somewhat differently and need different approaches for their modelling and analysis. Note that in the case of fork/join nodes with a synchronising queue, there will be one such queue associated with each of the siblings. This will essentially be a buffer holding a sub-job until it can be recombined with the sub-jobs from the other sibling queues

If there is no synchronizing queue, as in Figure 6.5, then a sub-job finishing service at a sibling queue will be forced to wait at that queue itself, blocking the server there for other customers in that sibling queue. The servers at all the sibling queues are released simultaneously when all the sub-jobs finish their required service at all the siblings. When that happens, the sub-jobs are recombined and the resultant job is released to depart from the fork/join node. Simultaneously, the servers at the sibling queues become free and can now start serving the sub-jobs for the next job, if any, entering the fork/join node. Note that effectively the service time for a job in the fork/join node will be the maximum of the service times of the individual sub-jobs and that the fork/join node cannot serve the sub-jobs of any subsequent job until the previous job leaves the node.

On the other hand, if a synchronizing queue is present, as in Figure 6.6, then a completed sub-job will move to the synchronising queue of that sibling queue after it gets the desired service. This will free the server at that sibling queue for the sub-jobs of subsequent jobs entering the fork/join node. When all the sub-jobs of a particular job are present in their respective synchronising queues, the fork/join node will combine them into a single job and will let it depart from the node.

Note that in both cases, the number of sub-jobs that is generated by a job entering the fork/join node is always equal to the number of sibling queues and these always get recombined into a single job once again before leaving the fork/join node. In general, the sibling queues inside the fork/join node may be of different types and an exact solution of this node would be difficult. We make the simplifying assumption that *all the sibling queues are single-server FCFS queues with exponentially distributed (possibly different) service times*. Some approximate models to analyse a fork/join node in open or closed queueing networks of infinite capacity queues are given next.

6.3.1 Fork/Join Node without Synchronising Queues in Open or Closed Networks of Infinite Capacity Queues

In this case, the service time encountered by a job entering the fork/join node will be the *maximum of the service times* for the sub-jobs at the k sibling queues. Assuming that the service provided at the sibling queues are independent of each other, the probability density function of this service time (i.e. of the fork/join node) may be found from the probability density function/cumulative distribution function of the service times of the k sibling queues. For doing this, we can use the result that if X and Y are independent random variables and Z is a random variable defined as $Z=max(X,Y)$, then the cumulative distribution function $F_Z(z)$ and probability density function $f_Z(z)$ of Z are respectively given by

$$F_Z(z) = F_X(z)F_Y(z) \tag{6.39}$$

$$f_Z(z) = f_X(z)F_Y(z) + f_Y(z)F_X(z) \tag{6.40}$$

where $F_X(x)$, $F_Y(y)$, $f_X(x)$ and $f_Y(y)$ are the cumulative distribution function and probability density function of X and Y, respectively.

Note that the probability density function and cumulative distribution function of Z may be found using Eqs. (6.39) and (6.40), if the probability density function and cumulative distribution function of X and Y are known. This result may also be easily extended for the case of k random variables corresponding to the service times of the k sibling queues. One way to do this will be to take the maximum of any two random variables and then take the maximum of this result with the next random variable and continue this until all the variables have been considered. Given the service time distributions of the individual (single server) sibling queues, the overall service time distribution of a job entering the fork/join node can then always be found. We can then also use this service time distribution to find the mean and SQV of the job's service time in a fork/join node of this type.

Even though the above calculation of the service time distribution can be done for any given sub-job service time distribution at the sibling queues, the results are greatly simplified if we assume the sub-jobs to be independent, exponentially distributed random variables. In this case, let $1/\mu_i$ be the mean of the (exponentially distributed) service time of a sub-job at the ith sibling queue, $i=1,.......,k$. Let X be the random variable denoting the overall service time of a job at the fork/join node without synchronising queues. Using the earlier approach and Eq. (6.39), we can then write the cumulative distribution function $F_X(x)$ of the overall service time at this fork/join node as

$$F_X(x) = \prod_{i=1}^{k} \left(1 - \exp(-\mu_i x)\right) \tag{6.41}$$

Open Network

In this case, we consider the situations where there are fork/join nodes of this type in an open network of GI/G/m queues or if such nodes are being considered in isolation. We can then find the cumulative distribution function and probability density function of the overall service times at the fork/join nodes; we can use Eq. (6.41) directly if the sub-jobs have independent, exponentially distributed service times. We can then use these distributions to find the mean and SQV of the resultant overall service time random variable at each of the fork/join nodes. Using these two service parameters, the approach of Section 6.2 may be directly applied to obtain the required solution.

Closed Network

In this case, we consider the situation where there are fork/join nodes of this type in a closed network. For analysing such a network, we would like to apply the MVA or the Convolution Algorithms. This however requires that the service times at all the queues should be exponentially distributed in nature. To approximately satisfy this condition, we fit an exponential distribution to the resultant distribution of the overall service time at each of the fork/join nodes. Note that the resultant distributions are the ones obtained as the distribution of the maximum of the k random service times at each of the k sibling queues of a fork/join node. This may be done by simply matching the first moments, as an exponential distribution is completely characterised by its mean. However, simulations show that a somewhat better way is to minimise the mean square error between the two distributions to get the best exponential fit. For this, let $F_X(x)$ be the resultant cumulative distribution function of the service time X at the fork/join node. (Note that if the sibling queues have exponentially distributed service times then this will be given by Eq. (6.41).) Let $F_{estimated}(x)$ be the cumulative distribution function (to be found) of the exponentially distributed minimum mean square error fit to $F_X(x)$. This may then be found as

$$F_{estimated}(x) = 1 - e^{-\hat{U}x} \qquad \text{exponential fit to } F_X(x) \tag{6.42}$$

$$\varepsilon(x,\hat{U}) = \int_0^\infty [F_X(x) - F_{estimated}(x)]^2 \, dx \qquad \text{mean square error} \tag{6.43}$$

$$\hat{U}_{\min} = \min_{\hat{U}}[\varepsilon(x,\hat{U})] \qquad \text{minimise the mean square error} \tag{6.44}$$

This leads to

$$F_{estimated}(x) = 1 - e^{-\hat{U}_{\min}x} \tag{6.45}$$

as the desired distribution. Once these resultant approximate exponential service time distributions are found for each of the fork/join nodes without synchronisation queues, we can use standard MVA or Convolution Algorithms to solve the overall queueing network in the same manner as described in Sections. 5.6 and 5.7.

6.3.2 Fork/Join Node with Synchronising Queues in Closed Networks of Infinite Capacity Queues

In this case, we cannot give a solution method (not even an approximate one) for solving an open network, which has one or more such fork/join nodes. However, Liu and Perros have proposed a method [LiP91] for solving a closed network of this kind. This is the method described here.

The algorithm proposed in [LiP91] for solving such a network involves an iterative approximation procedure. Consider a k-sibling fork/join queue with M customers. We first obtain a two-sibling closed fork/join queue by choosing any two of the siblings from the original k. This fork/join queue with j customers can be solved to obtain the throughput for $j=1,......,M$. This gives the Norton equivalent for the two-sibling fork/join queue. Another two-sibling queue is now obtained in which one of the siblings is the Norton equivalent from above and the other is one of the remaining siblings in the original k-sibling queue. This can be solved as before and the process repeated till all the k siblings have been used to obtain the Norton equivalent of the original k-sibling fork/join queue.

Consider a closed queueing network with M customers and one k-sibling, fork-join queue. Let A be the point of forking and B the point where the jobs are combined before leaving the system. To apply Norton's theorem between points A and B, consider first only the sibling queues 1 and 2 where the points A and B are shorted together. This system can be solved exactly as follows.

Let $P_{i,j}$ be the probability that there are i jobs in sibling queue 1 and j jobs in sibling queue 2. This is a closed queueing network with state-dependent service rates at the queues. Let (i, j) represent the state of the two queue system with i customers in sibling queue 1 and j customers in sibling queue 2 with probability $P_{i,j}$. Note that all combinations of i and j are not feasible. For example, if $M=3$, then the only feasible states are $(0,3)$, $(1,3)$, $(2,3)$, $(3,3)$, $(3,2)$, $(3,1)$ and $(3,0)$. It can be shown that for a general value of M, the total number of feasible states are $[(M+1)^2 - M^2]$. Let $\mu_1(i)$ be the service rate of sibling queue 1 when it has i sub-jobs and let $\mu_2(j)$ be the service rate of queue 2 when it has j sub-jobs. We can then draw the corresponding state transition diagram and obtain the following balance equations

$$P_{M,M}[\mu_1(M)+\mu_2(M)] = P_{M,M-1}\mu_1(M) + P_{M-1,M}\,\mu_2(M)$$
$$P_{M,n}[\mu_1(M)+\mu_2(n)] = P_{M,n-1}\mu_1(M) + P_{M,n+1}\mu_2(n+1) \qquad 1 \le n \le M-1$$
$$P_{M,0}\,\mu_1(M) = P_{M,1}\,\mu_2(1)$$
$$P_{n,M}[\mu_1(n)+\mu_2(M)] = P_{n+1,M}\,\mu_1(n+1) + P_{n-1,M}\,\mu_2(M) \qquad 1 \le n \le M-1$$
$$P_{0,M}\,\mu_2(M) = P_{1,M}\,\mu_1(1)$$
$$P_{M,M} + \sum_{i=0}^{M-1} P_{M,i} + \sum_{j=0}^{M-1} P_{j,M} = 1$$

Solving this set of equations, we can get the corresponding state probabilities for this two-sibling queue case as

$$P_{n,M} = P_{M,M}\prod_{j=n+1}^{M}\rho_2(j) \qquad 0 \le n \le M-1$$
$$P_{M,n} = P_{M,M}\prod_{j=n+1}^{M}\rho_1(j) \qquad 0 \le n \le M-1 \tag{6.46}$$
$$P_{M,M} = \frac{1}{1+\sum_{i=0}^{M-1}\prod_{j=i+1}^{M}\rho_2(j)+\sum_{k=0}^{M-1}\prod_{l=k+1}^{M}\rho_1(l)}$$

where

$$\rho_1(i) = \frac{\mu_2(i)}{\mu_1(M)} \tag{6.47}$$

and

$$\rho_2(j) = \frac{\mu_1(j)}{\mu_2(M)} \tag{6.48}$$

The system throughput of this two-sibling fork-join node with forking and joining nodes shorted together and M users is then given by

$$\hat{U}_{eff}(M) = \mu_2(M)\sum_{i=0}^{M-1} P_{i,M} + \mu_1(M)\sum_{j=0}^{M-1} P_{M,j} \tag{6.49}$$

This may be calculated for different M and these results may then be used to get the Norton's equivalent FES (flow equivalent server) for the two sibling queues for the steps of the subsequent reduction. In the next step, we use the Norton's equivalent FES for the first two sibling queues and combine this FES with the third queue, if any, to obtain an new FES for the new combination. This process is continued in steps, until all the k sibling queues have been combined to generate a single equivalent Norton's FES for the fork/join node. The equivalent FES obtained in this fashion then replaces the fork/join node in the actual network. If there are more than one such fork/join nodes, then each one of them is reduced to their equivalent FES in a similar manner. The resultant (closed) network is then solved by the usual methods (MVA or Convolution) to evaluate its overall performance and that of the individual nodes. A simpler approximation technique for this approach is also given in [LiP91].

We can also handle closed networks where some of the fork/join nodes have synchronising queues while the other fork/join nodes do not. In this case, the fork/join nodes with synchronising queues are replaced by their equivalent FES. The fork/join nodes without synchronising queues are replaced by suitably approximated single server queues with exponentially distributed service times. The resultant closed network can then be solved using the MVA or Convolution algorithms. Unfortunately, this option of mixing the types of fork/join nodes cannot be handled analytically in an open network. This is because, for open networks, we do not have a suitable approximation method for handling fork/join nodes with synchronising queues.

6.4 Models of Blocking in Open and Closed Networks of Finite Capacity Queues

Consider the queueing network scenario shown in Figure 6.7 where we now assume that the queues (at least the queues Q_j and Q_k) are of finite capacity. To illustrate blocking, consider the situation when a job finishes service at Q_i and wants to move to Q_j (with probability p_{ij}) or to Q_k (with probability p_{ik}). No blocking will occur if the buffer of the target queue has enough capacity to store the job. However, if the buffer of the target queue is full, the job will encounter blocking and cannot be expected to move to the target queue as required.

In general, blocking may arise in a network of queues where some or all queues have finite buffer capacities. The flow of customers from a node will be blocked if the corresponding destination node is full, i.e. all its servers are busy and all waiting positions are full with waiting customers. Blocking in such a network will be handled depending on the blocking mechanism that is being adopted. A blocking mechanism is a set of rules specifying when a node is blocked, what happens during the blocking period, and how a node becomes unblocked. Note that deadlock conditions may also arise in such networks. Various types of blocking mechanisms have been considered in the literature. Some of the more commonly identified blocking mechanisms have been summarised next. More details may be found in [Onu90], [Per89] and [Per94].

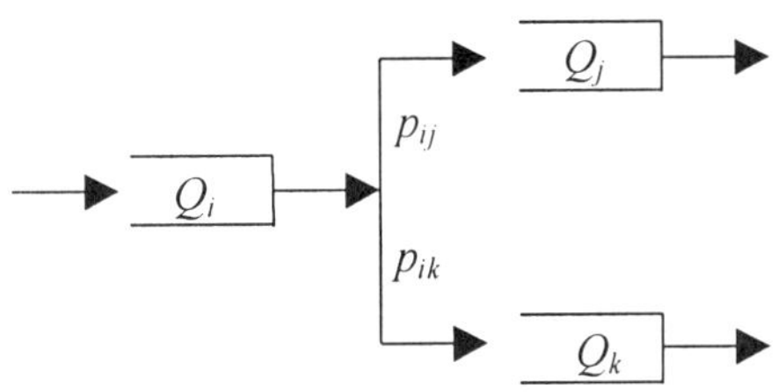

Figure 6.7. Blocking in a Queueing Network with Finite Capacity Queues

6.4.1 Rejection Blocking

If the customer is forced to leave the network as soon as it encounters a full node as its destination node, then the blocking mechanism is called Rejection Blocking. This kind of blocking essentially models loss systems and may be used in open networks of finite capacity queues. Since customers who are blocked are made to leave the system, this kind of blocking mechanism is not relevant for closed networks. This model has been widely used to model computer systems and packet-switching networks where jobs/packets are lost if they cannot be provided the desired buffer space. Deadlock conditions do not arise for this type of blocking as a blocked customer is always made to leave the system altogether.

6.4.2 Transfer Blocking

Consider a queueing network where a job on completion of its service at a source node attempts to join a destination node that is filled to its capacity. If the Transfer Blocking (or Blocking-After-Service, BAS) mechanism is being followed, the job waits at the source node blocking the server there, until it can enter the destination node. Note that it can do that only after a service completion at the destination node. In this blocking mechanism, therefore, the server at the source node becomes unblocked only when the number of customers in the destination node drops below its maximum capacity. This blocking mechanism has been used to model systems such as disk I/O configurations and manufacturing systems. In queueing networks with arbitrary topologies, it is possible that a node blocks more than one customer at the same time. Normally, a First-Blocked-First-Served rule is used to handle such a situation, i.e. the customer that was blocked first will also be the first to get unblocked.

Note that deadlock is possible in transfer blocking networks. If in a network, all the stations in any directed cycle are full at the same time and a blocked customer is scheduled to go to the next station in the cycle, then the network gets deadlocked. Deadlock may be handled by including some kind of deadlock handling strategy into the model. A modelling assumption that is frequently made is that a deadlock, when it occurs, is immediately resolved by simultaneously moving all the blocked jobs to their respective destinations. This would then remove the deadlock immediately, whenever it occurs. Another approach would be to restrict the system to situations where deadlocks are impossible. In a closed network, this may be ensured if the total number of customers in the system is less than the total capacity of any of the directed cycles in the network.

6.4.3 Repetitive Service - Random Destination or Fixed Destination

The blocking model in which a customer immediately receives another service at the source node itself if its destination node is full (and keeps repeating this until the customer completes service at a time instant when the destination node is not full) is called Repetitive Service (RS) Blocking. In this type of blocking, two cases may arise - Repetitive Service with Fixed Destination (RS-FD) or Repetitive Service with Random Destination (RS-RD). In the first case, the customer attempts to join the same destination node after each repeated service completion as mentioned above. In the second case, each time the customer completes service, a new destination node is chosen independent of the previous choice; this choice is made according to the routing probabilities from the source node. A repetitive service mechanism with fixed destination has been used in modelling telecommunication systems. Similarly, a repetitive service mechanism with random destination has been used to model a flexible manufacturing system. Due to the inherent nature of this blocking mechanism, deadlocks may arise in both open and closed networks with RS-FD type of blocking. On the other hand, open networks with RS-RD blocking can never encounter deadlock, even though such deadlocks may happen in closed networks of this type.

6.4.4 Blocking-Before-Service

Another type of blocking which has been identified in some systems, such as communication networks, is Blocking-Before-Service. In this type of blocking, a customer declares its destination node, say node j, prior to starting service at the source node i. If the destination node j is full at that instant, the server of node i becomes blocked, i.e. it cannot serve other customers. When a departure occurs from destination node j, the server at node i becomes unblocked and the customer begins receiving service.

In the subsequent Section 6.5 and 6.6., we present some approximate methods which may be used to analyse networks of finite capacity queues with *Rejection Blocking, Transfer Blocking* and *Repetitive Service Blocking* (with either *Fixed Destination* or *Random Destination*) as the blocking mechanisms being followed. We consider separately the methods used for open and closed networks of such queues. In general, we assume that all the nodes of the network are of finite capacity. However, the methods given here may generally be extended for networks where some of the queues are of finite capacity while the others are infinite capacity queues. This can typically be done by assuming sufficiently large (i.e. effectively infinite) buffer sizes for the nodes that are to be modelled as infinite capacity queues. For simplicity, we also assume here that the blocking mechanism is the same every where in the network. Modelling and analysing networks with different blocking mechanisms at different places in the network will be considerably more complex in nature.

6.5 Approximate Analytical Methods for Solving Closed Networks of Finite Capacity Queues

We present here methods for solving closed networks of finite capacity queues under conditions of *Transfer Blocking, Repetitive Service with Random Destinations* (RS-RD) and *Repetitive Service with Fixed Destination* (RS-FD) *Blocking*. Since the network is closed, no loss of jobs can occur and hence the *Rejection Blocking* model will not be relevant for such a network.

6.5.1 Transfer Blocking

Consider a situation where a job on completing service at node *j* attempts to join node *k*. In this case, blocking will arise if node *k* is full at that instant. In the case of Transfer Blocking (BAS or Blocking After Service), the blocked job at node *j* waits at the server (keeping it blocked) until it can enter node *k*. A product-form approximation method has been proposed by Akyilidiz [Aky89] for the analysis of Closed Queueing Networks with Transfer Blocking. This technique assumes the use of deadlock avoidance so that a deadlock cannot happen in the system being analysed. In order to ensure this, it is assumed that the number of customers in the system is less than the total capacity of that directed cycle in the network, which has the minimum total capacity. This assumption therefore implies that no directed cycle can ever have all its stations full at the same time and deadlocks will therefore be impossible. This technique seems to work well for various networks but its results cannot really be guaranteed to be the correct ones in all cases.

To analyse a closed queueing network with transfer blocking, this approach first considers an identical network where there is no blocking, i.e. one where all the nodes have infinite capacity. All the states of the equivalent non-blocking network will not be feasible in the blocking network. The states violating the constraints on the capacities of the various nodes are appropriately "normalised". This "normalisation" procedure is done in the following fashion.

If the number of customers in a node exceeds its capacity (before normalisation), then the number in that node is set equal to the capacity of the node and the excess jobs are distributed among all the other nodes depending on the incoming transition probabilities (i.e. those from the rest of the nodes to the node whose capacity constraints were being violated). We continue to do this until a "feasible" state is obtained. A "feasible" state is one that is a valid state for the blocking network. By normalising all the infeasible states of the identical non-blocking network, we obtain an equivalent state space that consists only of those states that are feasible for the blocking network. The probability P_H of a feasible state 'H' is the sum of the probabilities of all the states of the equivalent non-blocking network, which get normalised to this feasible state. This has been illustrated in Figure 6.8. For example, in this figure, the states 'a' and 'b' in the non-blocking network both reduce to state 'x' in the actual blocking network after the normalisation process. Similarly, states 'c' and 'd' of the non-blocking network correspond to states 'y' and 'z', respectively, in the blocking network after normalisation. It should be noted that this normalisation procedure is effectively like an enumeration of the states to identify the correspondence between one or more states in the non-blocking network with states in the blocking network. This would be difficult to do for large (closed) queueing networks or for networks with a large number of circulating jobs. This limits the applicability of this approximate analytical technique to small and medium sized networks.

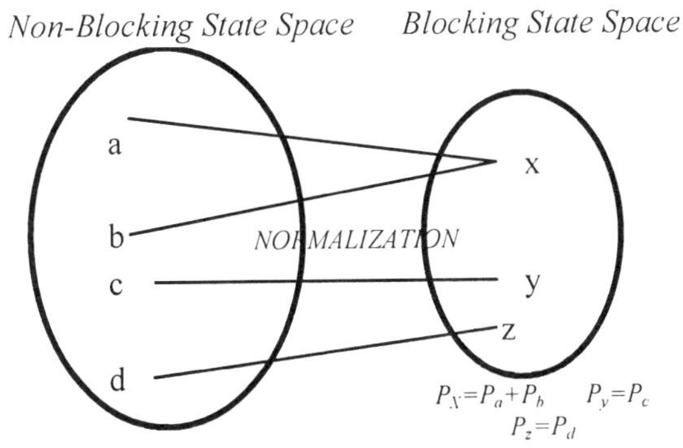

Figure 6.8. Mapping the State Space for Closed Network with Transfer Blocking

Note that for the non-blocking network, a product-form solution may be used to calculate its state probabilities. Based on the normalisation described above, the state distribution of the blocking network may be found from that of the non-blocking network. All the parameters, except throughput, can now be calculated from this state distribution. Calculation of the throughput is based on the fact that a non-blocking network with an equal number of states as the blocking network has the same stochastic structure as that of the blocking network. The throughput of these networks will also be approximately the same. In order to compute the throughput, we determine the number of states in the blocking queueing network. The equivalent number of jobs in the non-blocking network is then obtained such that the two networks have the same number of states. The non-blocking queueing network with this equivalent number of jobs is now analysed to obtain its throughput. This will also be the approximate throughput of the blocking network. The details of this approximate analysis algorithm are given next. We summarise below the assumptions made in this approach for analysing a general, closed network of multi-server finite-capacity queues with transfer blocking.

- The number of customers in the network is fixed and all customers belong to the same class.
- Service times at all the nodes are exponentially distributed.

- FCFS service discipline is followed at all the nodes.
- The transition probability matrix is static.
- In situations where more than one customer is blocked by the same node, the blocked customers enter the node on a first-blocked-first-enter basis.
- Some or all the nodes in the network may have infinite capacity.
- The network is deadlock free.

The following general notations are used in the algorithm given subsequently.

N	number of nodes in the network
μ_i	service rate at node i
m_i	number of servers at node i for $i = 1...N$
V_i	visit Ratio of ith node with respect to node 1
M_i	capacity of the ith node (number of servers and the number of waiting positions)
[*P*]	routing probability matrix with p_{ij}, $i,j=1,......N$, as the probability of a job finishing service at node i getting routed to node j
K	number of jobs/customers in the (closed) network

The actual algorithm is given below.

Algorithm

Step 1. Solve for the visit ratios using

$$V_i = \sum_{j=1}^{N} V_j\, p_{ji} \qquad i = 1...N \tag{6.50}$$

for V_i, the mean number of visits to node i, assuming node 1 to be the reference node, i.e. $V_1=1$.

Step 2. Find state probabilities using

$$\begin{aligned} p^*(\boldsymbol{k}^*) &= \frac{1}{G(K)} \prod_{i=1}^{N} \left[\frac{V_i}{\mu_i \beta_i(k_i^*)} \right]^{k_i^*} \\ \beta_i(k_i^*) &= k_i^*! \qquad\qquad k_i^* \le m_i \\ &= m_i!(m_i)^{k_i^* - m_i} \qquad k_i^* > m_i \end{aligned} \tag{6.51}$$

where $G(K)$ is the normalisation constant and $p(\boldsymbol{k}^*)$ is the equilibrium state probability distribution of the network without any capacity restrictions, i.e. the equivalent non-blocking network. This may be found using any of the typical methods described in Section 5.6 and 5.7.

Step 3. Normalise the customers using the following redistribution approach

For each i, $i=1,\ldots\ldots,N$, if $k_i^* > M_i$, then set k_j^* $j=1,\ldots\ldots,N$ as follows to normalise the customers

$$\begin{aligned} k_j^* &= M_j && j=i \\ &= k_j^* + (k_i^* - M_i)\frac{V_j p_{ji}}{V_i(1-p_{ii})} && j \neq i \end{aligned} \tag{6.52}$$

until $k_i^* \le M_i$, i.e. there is no violation of the capacity constraint of node i.

Step 4. The equilibrium state probability distribution for the feasible states is computed as -

$$p(\boldsymbol{k}) = \sum_{\boldsymbol{k}^* \text{ where } f(\boldsymbol{k}^*)=\boldsymbol{k}} p^*(\boldsymbol{k}^*) \tag{6.53}$$

where $(\boldsymbol{k})$ is the "normalised" state or state corresponding to the blocking network and the function f transforms the non-feasible state $(\boldsymbol{k}^*)$ to the feasible state $(\boldsymbol{k})$.

Step 5. Calculate the number of states $Z'(K)$ of the blocking network as $Z' = Z_1 \otimes Z_2 \otimes \ldots\ldots\ldots\ldots \otimes Z_N$ where $\otimes$ is the convolution operator and Z_i $i=1,2,\ldots\ldots\ldots,N$ is a $(K+1)$ dimensional vector given by

$Z_i=[z_i(0), z_i(1)\ldots\ldots\ldots\ldots\ldots\ldots\ldots\ldots z_i(K)]^T$ with

$$z_i(k) = 1 \qquad \text{if } k = 0,1,\ldots\ldots\ldots, M_i + \sum_{j=1 \text{ and } p_{ji}>0}^{N} m_j$$
$$= 0 \qquad \text{otherwise}$$

Note that in the same network, if there are no capacity limitations, then the number of states will be given by the binomial expression

$$Z = {}^{N+K-1}C_{N-1}$$

Step 6. Calculate the equivalent number of customers K_e such that

$$Z_e = {}^{N+K_e-1}C_{N-1}$$

is approximately equal to $Z'(K)$.

Step 7. Using any Product Form Algorithm such as MVA or Convolution, analyse the non-blocking network with K_e total number of customers to obtain the total throughput $\lambda_{NB}(K_e)$.
As explained earlier, this will also be approximately the throughput $\lambda_B(K)$ of the actual blocking network with K customers.

$$\lambda_B(K) = \lambda_{NB}(K_e) \tag{6.54}$$

Step 8. The throughput λ_i and utilisation ρ_i of each node i, $i=1,\ldots\ldots,N$ are computed as

$$\lambda_i(K) = V_i \lambda_B(K) \qquad\qquad \rho_i(K) = \frac{\lambda_i(K)}{m_i \mu_i} \tag{6.55}$$

Step 9. The mean time taken by a job to circulate once through the whole system is referred to as the mean total response time or mean cycle time T_C of the closed network. This may be computed from

$$T_C(K) = \frac{K}{\lambda_B(K)} \tag{6.56}$$

Step 10. Using the state distribution obtained in Eq. (6.53) after the normalisation procedure, we can calculate the mean number $N_i(K)$

of jobs at node i, when there are K jobs circulating in the closed network. Using this, the mean time T_{iC} spent by a job at node i in one cycle, will be

$$T_{iC}(K) = \frac{N_i(K)}{\lambda_B(K)} \tag{6.57}$$

Note that, this algorithm ends up by providing all the required node and network parameters that will generally be of interest. For the network as a whole, it provides the network throughput $\lambda_B(K)$, the cycle time $T_C(K)$ and the approximate state distribution vector $p(\mathbf{k})$ for K circulating jobs in the network. For each individual node, say node i, it provides the visit ratio V_i, the throughput $\lambda_i(K)$, the node utilisation $\rho_i(K)$, the mean number in the node $N_i(K)$, and the mean time $T_{iC}(K)$ spent in the node by a job in one cycle.

6.5.2 Repetitive Service (RS) Blocking

Consider a situation where a customer upon completion of its service at node i, attempts to join node j. If node j is full, the customer immediately receives another service at node i. This will continue until the customer can actually move to node j on its service completion. This type of blocking is termed Repetitive Service Blocking. In the above description, we assume that the destination node is fixed. This is essentially *Repetitive Service Blocking with Fixed Destination* (RS-FD). Alternatively, if the customer chooses a destination node (from the set of nodes allowed by the routing probabilities) independently after every service completion, and tries to move to it then the blocking mechanism is called *Repetitive Service with Random Destination* (RS-RD). We give here the Maximum Entropy Method of Kouvastos et al [KoX89] for the analysis of arbitrary queueing networks with multiple general servers and Repetitive Service blocking.

The Maximum Entropy technique is based on the principle that "of all the distributions satisfying the set of constraints imposed by the system, the most likely distribution is the one that maximises the system's entropy". In the analysis of closed networks with Repetitive Service Blocking, a related "pseudo-open" network is first solved. This "pseudo-open" network is a closed network, which is represented as an open network with no external arrival streams and no external departures. The pseudo-open network must also satisfy the principle of "conservation of population". This is represented by the fixed mean population constraint, i.e., the mean number of jobs at all the nodes in network must sum up to the number of jobs actually circulating in the closed network. The individual queues in the closed network are "censored", i.e. arrivals to a queue when all the N buffers are full are turned away and, moreover, no departures are allowed to occur when the queue has minimum possible number of jobs K ($K \geq 0$). In closed networks, K will be greater than zero for some node i if the total number of jobs circulating in the network is greater than the total capacity of the network, excluding this node. If L is the number of jobs in the network, a node with capacity N greater than or equal to L will effectively behave as an infinite capacity node. In this case, the capacity of such a node (having $N \geq L$) can instead be set to the number of jobs L without making any difference in the analysis. The pseudo-open network is decomposed into individual censored queues with revised inter-arrival and service time distributions and solved in the same way as an open network under Repetitive Service Blocking. The analysis of open queueing Networks with Repetitive Service Blocking has been explained subsequently in Section 6.6.2. The solution obtained from the pseudo-open network may not satisfy the job flow balance equations. The values of the flow balance coefficients are then iteratively adjusted so that the flow balance equations are satisfied. The marginal state probabilities obtained after these iterations may now be used to obtain the performance measures of interest. It is, however, suggested that an efficient technique be used to compute the normalisation constant for the calculation of the state probabilities. An iterative convolution method has been proposed as a suitable method to calculate this normalisation constant.

The details of the analytical approach will be beyond the scope of this text. For this, the reader is referred to the original papers by Kouvastos et al., such as [KoX89]. We give the algorithm here in some detail as it may be of use to obtain the performance parameters of a closed network with RS-RD or RS-FD blocking. We summarise below the assumptions made in this approach for the analysis of a general closed network of finite capacity queues with multiple general servers and Repetitive Service blocking.

- The number of customers in the network is fixed and all customers belong to the same class.
- A First-Come-First-Served (FCFS) service discipline is followed at each node.
- The routing probability matrix is static.
- Some or all the nodes in the network may have infinite capacity.
- Each queue in the network is modelled as a GE/GE/c/K;N ($0 \leq K \leq N$) censored queue with Generalised Exponential (GE) inter-arrival and service times, where c is the number of servers, N is the capacity of the queue and K is the censoring value mentioned earlier, (i.e. departures not allowed from state K).

The following notations are used in the subsequent algorithm -

- M number of nodes in the network
- L number of customers in the network
- c_i number of servers at node i, $c_i \geq 1$
- N_i capacity of node i (including the number of servers)
- μ_i service rate at node i
- C_{Si}^2 SQV of the service time at node i
- K_i minimum number of customers always present at node i
- L_i virtual capacity of node i, i.e. $L_i = Min(N_i, L)$
- λ_{0i} the overall external arrival rate to node i; this will be zero for a closed network
- C_{0i}^2 SQV of the external inter-arrival times
- p_{ij} routing probability that a job completing service at node i attempts to join node j
- π_{ij} probability that a job completing service at node i is blocked by node j
- π_{0i} probability that an external arrival finds node i in state N_i
- π_{di} probability that a job completing service at node i is blocked under the RS-RD blocking mechanism

The actual algorithm for both the RS-RD and the RS-FD mechanisms is given below.

Algorithm

Stage 1: Solving the Pseudo-Open Network

Step 0. Remove immediate feedback for all nodes $i = 1, \ldots\ldots, M$.
If $p_{ii} > 0$, then this is done as follows -
$\mu_i = \mu_i\,(1 - p_{ii}), \quad C_{Si}^2 = p_{ii} + (1-p_{ii})C_{Si}^2$
$p_{ij} = p_{ij}/(1-p_{ii}) \quad$ *for* $i \neq j, j=1,2,\ldots\ldots,M$ and $p_{ij} = 0$ *for* $i=j$
Note that this procedure is the same as described to remove immediate feedback in Section 6.2.

Step 1. Find L_i and K_i using the following -

$$L_i = \min(N_i, L)$$
$$K_i = \max(0, L - \sum_{j \neq i, j=1}^{M} N_j) \tag{6.58}$$

Step 2. Solve the following flow balance equations for the initial values of the individual flows to each queue in the network

$$\tilde{\lambda}_i = \sum_{j=1}^{M} \hat{p}_{ji} \tilde{\lambda}_j \qquad i=1,\dots\dots,M \tag{6.59}$$

The effective routing probabilities $\hat{p}_{ji}$ are computed iteratively later but the iterations may be started with $\hat{p}_{ji} = p_{ji}$

Step 3. Solve iteratively, the set of non-linear equations given below -

$$\pi_{ij} = \sum_{n_j=0}^{c_j-1} p_i(n_j)(1-\tau_{ij})^{N_j-n_j} \left(\frac{\sigma_j}{\sigma_j(1-\tau_{ij})+\tau_{ij}} \right)^{c_j-n_j} + \sum_{n_j=m_j}^{N_j} p_j(n_j)(1-\tau_{ij})^{N_j-n_j} \tag{6.60}$$

$$\sigma_j = \frac{2}{1+\tilde{C}_{Sj}^2} \qquad j=1,\dots\dots\dots,M \tag{6.61}$$

$$\tau_{ij} = \begin{cases} \dfrac{2}{1+C_{dij}^2} & i \neq 0 \\ \dfrac{2}{1+C_{0i}^2} & i = 0 \end{cases} \qquad i,j=1,\dots\dots\dots,M \quad i \neq j \tag{6.62}$$

$$L = \sum_{i=1}^{M} E\{n_i\} \text{ where } E\{n_i\} \text{ is the mean number of jobs in node } i \tag{6.63}$$

$$\pi_{di} = \sum_{j=1}^{M} \hat{p}_{ij} \pi_{ij} \tag{6.64}$$

$$\widetilde{\lambda}_{0i} = \lambda_{0i}(1 - \pi_{0i}) \qquad \widetilde{C}_{0i}^{2} = \pi_{0i} + (1 - \pi_{0i})C_{0i}^{2} \tag{6.65}$$

$$\widetilde{\mu}_{i} = (1 - \pi_{di})\mu_{i} \qquad \widetilde{C}_{Si}^{2} = \pi_{di} + (1 - \pi_{di})C_{Si}^{2} \qquad \text{RS-RD} \tag{6.66}$$

$$\widetilde{\mu}_{i} = \frac{\mu_{i}}{\displaystyle\sum_{j,\, p_{ij}>0} \frac{p_{ij}}{1 - \pi_{ij}}}$$

$$\widetilde{C}_{Si}^{2} = -1 + \frac{C_{Si}^{2}}{\displaystyle\sum_{j,\, p_{ij}>0} \frac{p_{ij}}{1 - \pi_{ij}}} + \frac{\displaystyle\sum_{j,\, p_{ij}>0} \frac{p_{ij}(1 + \pi_{ij})}{(1 - \pi_{ij})^{2}}}{\left(\displaystyle\sum_{j,\, p_{ij}>0} \frac{p_{ij}}{1 - \pi_{ij}}\right)^{2}} \qquad \text{RS-FD} \tag{6.67}$$

$$\lambda_{i} = \frac{\widetilde{\lambda}_{i}}{1 - \pi_{i}} \qquad C_{i}^{2} = \frac{\widetilde{C}_{i}^{2} - \pi_{i}}{1 - \pi_{i}} \tag{6.68}$$

$$\pi_{i} = \frac{\lambda_{0i}\pi_{0i} + \displaystyle\sum_{j=1}^{M} \lambda_{j} \hat{p}_{ji} \frac{\pi_{ji}}{(1 - \pi_{ji})}}{\lambda_{0i} + \displaystyle\sum_{j=1}^{M} \lambda_{j} \hat{p}_{ji} \frac{1}{(1 - \pi_{ji})}} \tag{6.69}$$

$$\widetilde{C}_{dj}^{2} = \widetilde{\rho}_{j}^{2} \widetilde{C}_{Sj}^{2} + (1 - \widetilde{\rho}_{j})\widetilde{C}_{j}^{2} + (1 - \widetilde{\rho}_{j})\widetilde{\rho}_{j} \qquad C_{dj}^{2} = \frac{\widetilde{C}_{dj}^{2} - \pi_{dj}}{1 - \pi_{dj}} \tag{6.70}$$

$$\frac{1}{1+\widetilde{C}_i^2} = \frac{\lambda_{0i}}{\lambda_i}\frac{1}{1+\widetilde{C}_{0i}^2} + \sum_{j=1}^{M}\frac{\lambda_j \hat{p}_{ji}}{\lambda_i}\frac{1}{1+\widetilde{C}_{dji}^2} \tag{6.70}$$

$$\text{with } \hat{p}_{ij} = \begin{cases} p_{ij}\dfrac{1-\pi_{ij}}{1-\pi_{di}} & RS-RD \\ p_{ij} & RS-FD \end{cases} \tag{6.80}$$

Details on the derivations of the above approximating formulae may be found in [KoX89].

The state distribution $p_i(n_j)$ for each queue $i=1,........,M$ is found by solving the censored GE(λ_i, C_i^2)/GE($\mu_i, \widetilde{C}_{Si}^2$)/$c_i$/$K_i$;$N_i$ queue using the maximum entropy based approximate method of Kouvastos. For a queue of this type, the results are

$$p(n) = p(K)G_n x^{h(n)} y^{f(n)} \qquad n=K+1,..........N \tag{6.81}$$

Note that since the queue is censored, departures cannot occur once the system state reaches K. Therefore, the system state cannot become less than K. Moreover, since the capacity is finite, i.e. N, the system state cannot exceed N. For evaluating the state probability using Eq. (6.81), we use the following.

λ = Average arrival rate of the arrival process to the queue

C^2 = SQV of the inter-arrival time of the arrival process to the queue

$$G_n = \prod_{l=K+1}^{m(n)} g(l) \quad J = \max(c, K+1) \quad h(n) = \max(0, n-J)$$
$$f(n) = \max(0, n-N+1) \quad m(n) = \max(K+1, \min(c, n)) \tag{6.82}$$

$$g(K+1)=\begin{cases}\dfrac{\tau c\rho\sigma}{(K+1)[\sigma(1-\tau)+\tau]} & K<c-1\\ \dfrac{\tau c\rho\sigma}{\tau\rho(1-\sigma)+\sigma} & K=c-1\\ \dfrac{\sigma(1-\tau)+\tau}{\tau\rho(1-\sigma)+\sigma} & K\geq c\end{cases} \tag{6.83}$$

$$g(l)=\begin{cases}\dfrac{\tau c\rho+(l-1)\sigma(1-\tau)}{l[\sigma(1-\tau)+\tau]} & l<J\\ \dfrac{\sigma[\tau c\rho+(J-1)\sigma(1-\tau)]}{J[\tau\rho(1-\sigma)+\sigma]} & l=J\end{cases} \qquad l=(K+2),\ldots\ldots,J \tag{6.84}$$

with

$$x=\frac{\tau\rho+\sigma(1-\tau)}{\tau\rho(1-\sigma)+\sigma}\quad y=\frac{1}{1-(1-\sigma)x}\quad \sigma=\frac{2}{1+C_S^2}\quad \tau=\frac{2}{1+C^2}\quad \rho=\frac{\lambda}{c\mu}$$

and $p(K)$ may be found by using the normalisation condition that

$$\sum_{n=K}^{N} p(n)=1 \tag{6.85}$$

The iterations of *Step 3* above are performed until we obtain sufficient convergence in the values of $\tilde{C}_{di}^2$ and $\tilde{C}_i^2$.

Stage 2: Solving the Closed Network

Step 4. Use the convolution algorithm of Section 5.6 to find the normalisation constant and use that to find the marginal state probabilities of the individual queues $\{p_i(n_j)\}$, $i=1,\ldots\ldots,M$.

Step 5. For each queue $i=1,\ldots\ldots,M$, the marginal mean queue length $E\{n_i\}$ and the mean throughput X_i may be calculated using

$$E\{n_i\} = \sum_{n_i=K_i}^{L_i} n_i p_i(n_i) \tag{6.86}$$

$$X_i = \tilde{\mu}_i \sum_{n_i=K_i+1}^{L_i} \delta_i(n_i) p_i(n_i) \tag{6.87}$$

where $$\delta_i(n_i) = \begin{cases} c_i[1-(1-\sigma_i)^{n_i-K_i}] & K_i \geq c_i \\ \min(n_i, c_i) & K_i < c_i \end{cases} \tag{6.88}$$

and $$\sigma_i = \frac{2}{1+C_{Si}^2} \tag{6.89}$$

Step 6. Using the value of y_i is the one obtained in *Step 3,* set

$$\tilde{y}_i = y_i \qquad i = 1, \ldots\ldots, M \tag{6.89}$$

and iteratively adjust the value of the flow balance coefficients by using the following -

$$y_i = \tilde{y}_i \frac{\tilde{\lambda}_i L}{\left(X_i \sum_{j=1}^{M} \frac{\tilde{\lambda}_j}{X_j} E\{n_j\} \right)} \tag{6.90}$$

Step 7. Repeat from *Step 4* for *i=1,.......,M* until we get $\dfrac{\lambda_i}{X_i} = \text{constant}$

6.6 Approximate Analytical Methods for Solving Open Networks of Finite Capacity Queues

We present here approximation techniques to handle open networks of finite capacity queues with a variety of blocking mechanisms. Rejection Blocking, Repetitive Service Blocking (both RS-RD and RS-FD) and Transfer Blocking have been considered.

6.6.1 Rejection Blocking

In Rejection Blocking, a customer arriving to a fully occupied node is lost, i.e. it is forced to leave the network. (This kind of blocking mechanism cannot therefore be considered in a closed network!). We give here a technique [BrG84] for solving open exponential queueing networks with Rejection Blocking, i.e. networks with Poisson arrival processes and exponentially distributed service times. This technique is based on the assumption that the net flow into the node, consisting of the external arrivals and the secondary flow from the other nodes in the network, may also be assumed to be Poisson in nature. Flow balance equations are solved iteratively in order to find the net offered traffic to node i. These equations are not the same as those for Jackson-type networks, but have been modified to incorporate Rejection Blocking. It is assumed that networks with Rejection Blocking may still be approximated by a product-form solution. Using this assumption, the network is decomposed into individual M/M/c/K queues and each queue is analysed independently, using the net offered traffic to this queue.

Apart from the average number, average sojourn time etc., parameters like conditional expected sojourn time for the customers who are eventually lost (U_l) or conditional expected sojourn time for the customers who left the network after completing their service requirements (U_s) will also be important for networks with Rejection Blocking. To calculate U_s and U_l, we need to consider the random walk of the customer through the network with "Loss" and "Served" as the absorbing states. These states respectively represent the situations when a customer was lost on encountering a full queue or when the customer left the network after completing its service requirements. Customers who are lost before starting service at any node in the network are referred to as *rejected customers*. These are essentially the external arrivals that encounter a full node upon entry into the network and are lost. On the other hand, a *lost customer* in this context is one who is lost before completing its service requirements. Lost customers also include those customers who are rejected right at their point of entry into the network. In this technique, we consider networks with exponentially distributed inter-arrival and service times. However, the nodes may have multiple servers. The assumptions inherent in this technique are summarised below.

- All external arrival processes are Poisson in nature.
- The service times at all the nodes are exponentially distributed.
- For each node, the superposition of the original Poisson flow from the external source with all the secondary flows coming

from various other nodes in the network is also assumed to be Poisson.
- The routing probability matrix is static.
- All the jobs in the network are of the same class.
- The service discipline at all the nodes in the network is FCFS.
- Some or all the nodes in the network may have infinite capacity.

The following notations are used in the subsequent algorithm -

M number of nodes in the network
m_i number of servers at node i
μ_i service rate at node i
λ_{0i} external arrival rate to node i
N_i capacity of node i (including the servers)

The actual algorithm to find approximate performance results for a network of this type is given below.

Algorithm

Step 0. Solve the following equations iteratively

$$\lambda_j = \sum_{i=1}^{M} \lambda_i^{out} p_{ij} + \lambda_{0j} \tag{6.91}$$

$$\lambda_i^{out} = \lambda_i \pi_i(\lambda_i) \tag{6.92}$$

$$\pi_i(\lambda_i) = \sum_{k=0}^{N_i-1} p_i(k) \tag{6.93}$$

$$p_i(k)=\begin{cases} p_i(0)\dfrac{\rho_i^k}{k!} & k \le m_i \\ p_i(0)\dfrac{\rho_i^k}{m_i!(m_i)^{k-m_i}} & m_i < k \le N_i \end{cases} \tag{6.94}$$

$$\rho_i = \frac{\lambda_i}{\mu_i} \qquad \sum_{k=0}^{N_i} p_k = 1$$

These iterations essentially solve the queueing network. Once sufficient convergence is obtained, the various output parameters of the queue may be found.

Step 1: Using the state distributions $p_i(k)$ for each queue in the system, $i=1,.....M,$ the various output parameters for the individual queues and the overall network are found as follows.

$$\text{Average number waiting in queue } i = N_{qi} = \sum_{k=m_i}^{N_i} (k - m_i) p_i(k) \tag{6.95}$$

$$\text{Therefore, } W_{qi} = \frac{N_{qi}}{\lambda_i^{out}} \quad W_i = W_{qi} + \frac{1}{\mu_i} \quad N_i = \lambda_i^{out} W_i \tag{6.96}$$

$$\text{Average Number of Busy Servers at queue } i = \frac{\lambda_i^{out}}{\mu_i} \tag{6.97}$$

The average total flow rate of customers completing their service in the network is -

$$\lambda_{out} = \sum_{i=1}^{M} \lambda_i^{out}\left(1 - \sum_{j=1}^{M} p_{ij}\right) \tag{6.98}$$

The average total flow rate of lost customers is

$$\lambda_{lost} = \lambda_0 - \lambda_{out} \tag{6.99}$$

where λ_0 is the average total flow of external arrivals to the network and will be given by -

$$\lambda_0 = \sum_{j=1}^{M} \lambda_{0j} \tag{6.100}$$

We can also find the average total flow rate of rejected customers - these are the external arrivals to the network which are rejected and lost as soon as they arrive at a node from outside. This will be given by -

$$\lambda_{rej} = \sum_{i=1}^{M} \lambda_{0i}[1 - \pi_i(\lambda_i)] \tag{6.101}$$

Note that the total flow rate of lost customers will also include the customers who are rejected. This is because the rejected customers are also counted as lost customers except that they get lost right on the first node where they enter the network. The average number of services received by a job entering the network will be given as

$$m = \sum_{i=1}^{M} \frac{\lambda_i^{out}}{\lambda_0} \tag{6.102}$$

and, similarly, the average number of services received by a customer who was not rejected, will be

$$\hat{m} = \sum_{i=1}^{M} \frac{\lambda_i^{out}}{\lambda_0 - \lambda_{rej}} \tag{6.103}$$

The sojourn times for a job entering the network are also of interest. We can define two sojourn times in this case. The average sojourn time of a job in the network will be

$$u = \sum_{i=1}^{M} \frac{\lambda_i^{out}}{\lambda_0}(W_{qi} + \frac{1}{\mu_i}) \tag{6.104}$$

This sojourn time will also include those customers (with sojourn time=0) which get rejected. We can define another average sojourn time for the customers who are not rejected, i.e. ones that actually do enter the network. (Even these customers may get lost later and may leave without getting the full service that they want.) The average sojourn time in the network for such customers, who are not rejected, will be

$$\hat{u} = \sum_{i=1}^{M} \frac{\lambda_i^{out}}{\lambda_0 - \lambda_{rej}} (W_{qi} + \frac{1}{\mu_i}) \tag{6.105}$$

The average conditional sojourn time U_l of a customer who was eventually lost from the network and the average conditional sojourn time U_s of a customer who actually completes its service are two other sojourn times that may be of interest. To compute these parameters, the approach taken is to consider the random walk of a customer in the network with the states L and S as the two absorbing states indicating the lost and served states, respectively. The lost state L is entered by a customer when it gets rejected at the entry to one of the queues in the network (the first queue or any subsequent queue) during its service sequence. The served state S is entered by a customer who gets all the desired service and leaves the queue at the completion of the full sequence of service that it requires. Note that we can represent the state of a customer to be either the queue in which it is currently present or the lost and served states, L and S. A customer entering the network will transit through one or more queues (i.e. states) but will eventually end up in one of the two absorbing states L and S. With suitable approximations to consider this as a Markov Chain, this random walk can be analysed to obtain the desired sojourn times U_l and U_s, mentioned earlier. The detail of this analysis are beyond the scope of this text but can be found in the original paper of Bronshtein and Gertsbakh.

6.6.2 Repetitive Service Blocking

Analysis of systems with Repetitive Service blocking has been described earlier for closed networks. In this section, we consider open networks with repetitive service blocking. The approximate analysis of this type of open queueing networks also uses the Maximum Entropy Method which was described earlier for the corresponding case of closed network.

Consider an arbitrary open queueing network under RS-RD or RS-FD blocking mechanisms. We assume that the network consists of M FCFS multiple server queues of finite or infinite capacity. It has been shown [KoX89] that the Maximum Entropy (ME) solution, subject to normalisation and certain other constraints, can give an approximate product form solution for the network of the type

$$p(\mathbf{n}) = p(n_1,....,n_M) = \prod_{i=1}^{M} p_i(n_i) \tag{6.106}$$

where $p(\mathbf{n})$ is the equilibrium state probability distribution of the network in state $\mathbf{n}$ and $p_i(n_i)$ is the probability that the ith queue is in state n_i. Following the MEM approach, $p_i(n_i)$ may be found as the marginal maximum entropy based solution for the corresponding GE/GE/c_i/N_i queue.

The open network to be analysed is decomposed into individual GE/GE/c_i/N_i queues with appropriate inter-arrival and service time distributions. All the incoming streams (including the external arrival stream) to node i merge to form the overall arrival stream to node i. The effective service time at node i is the total time for which a server of queue i is occupied by a particular job. The following assumptions are inherent in this approach and are summarised for convenience.

- A First-Come-First-Served (FCFS) service discipline is followed at each node.
- The routing probability matrix is static.
- Some or all the nodes in the network may have infinite capacity.
- Each queue in the network is modelled as a GE/GE/c/N queue with Generalised Exponential inter-arrival and service times

The following notations are used in the subsequent algorithm -

M number of nodes in the network
c_i number of homogeneous servers at node i, $c_i \geq 1$
N_i capacity of node i (including the number of servers)
μ_i service rate at node i
C_{Si}^2 SQV of the service time
λ_{0i} average external arrival rate to node i
C_{0i}^2 SQV of the inter-arrival times of the external arrivals
p_{ij} routing probability that a job completing service at node i attempts to join node j
π_{ij} probability that a job completing service at node i is blocked by node j.
π_{0i} probability that an external arrival finds node i in state N_i. This will also be the probability that an external arrival is blocked on entry at node i and is not allowed to enter the network
π_{di} probability that job completing service at node i is blocked under the RS-RD blocking mechanism

Algorithm

Step 0. Remove immediate feedback for all nodes $i = 1, ..., M$.
If $p_{ii} > 0$, then this is done as follows -
$\mu_i = \mu_i\ (1- p_{ii}), \quad C_{Si}^2 = p_{ii} + (1-p_{ii})C_{Si}^2$
$p_{ij} = p_{ij}/(1-p_{ii}) \quad for\ i \neq j, j=1,2,.....,M$ and $p_{ij} = 0\ for\ i=j$
Note that this procedure is the same as described to remove immediate feedback in Section 6.2.

Step 1. Initialisation for the subsequent iterations are done by choosing suitable initialising values for -

$\widetilde{C}_{di}^2$ and π_{di} for $i=1,......,M$

Note that any appropriate value may be chosen to start the subsequent iterations. The probability value should of course be chosen between *0* and *1* and we have got good results with a choice of *0.5*. The SQV value may be conveniently chosen as *1* which would actually be the case if the distribution was exponential in nature.

Step 2. Solve the following set of linear flow equations

$$\widetilde{\lambda}_i = \widetilde{\lambda}_{0i} + \sum_{j=1}^{M} \widetilde{\lambda}_j \hat{p}_{ji} \tag{6.107}$$

using

$$\widetilde{\lambda}_{0i} = \lambda_{0i}(1 - \pi_{0i}) \qquad i=1,.......,M \tag{6.108}$$

The effective routing probabilities $\hat{p}_{ji}$ are computed iteratively later but the iterations may be started with $\hat{p}_{ji} = p_{ji}$. The variable π_{0i} is also computed later but any suitable initial value can be used at this point.

Step 3. Solve iteratively the set of non-linear equations given below

$$\pi_{ij} = \sum_{n_j=0}^{c_j-1} p_i(n_j)(1-\tau_{ij})^{N_j-n_j}\left(\frac{\sigma_j}{\sigma_j(1-\tau_{ij})+\tau_{ij}}\right)^{c_j-n_j} + \sum_{n_j=m_j}^{N_j} p_j(n_j)(1-\tau_{ij})^{N_j-n_j} \tag{6.109}$$

$$\sigma_j = \frac{2}{1+\widetilde{C}_{Sj}^2} \qquad j=1,\dots\dots\dots,M \tag{6.110}$$

$$\tau_{ij} = \begin{cases} \dfrac{2}{1+C_{dij}^2} & i \neq 0 \\ \dfrac{2}{1+C_{0i}^2} & i = 0 \end{cases} \qquad i, j=1,\dots\dots\dots,M \quad i \neq j \tag{6.111}$$

$$\pi_{di} = \sum_{j=1}^{M} \hat{p}_{ij}\pi_{ij} \tag{6.112}$$

$$\begin{aligned} \widetilde{\lambda}_{0i} &= \lambda_{0i}(1-\pi_{0i}) \\ \widetilde{C}_{0i}^2 &= \pi_{0i} + (1-\pi_{0i})C_{0i}^2 \end{aligned} \tag{6.113}$$

$$\begin{aligned} \widetilde{\mu}_i &= (1-\pi_{di})\mu_i \\ \widetilde{C}_{Si}^2 &= \pi_{di} + (1-\pi_{di})C_{Si}^2 \end{aligned} \qquad \text{RS-RD} \tag{6.114}$$

$$\tilde{\mu}_i = \frac{\mu_i}{\sum\limits_{j,p_{ij}>0} \frac{p_{ij}}{1-\pi_{ij}}}$$

$$\tilde{C}_{Si}^2 = -1 + \frac{C_{Si}^2}{\sum\limits_{j,p_{ij}>0} \frac{p_{ij}}{1-\pi_{ij}}} + \frac{\sum\limits_{j,p_{ij}>0} \frac{p_{ij}(1+\pi_{ij})}{(1-\pi_{ij})^2}}{\left(\sum\limits_{j,p_{ij}>0} \frac{p_{ij}}{1-\pi_{ij}}\right)^2} \qquad \text{RS-FD} \quad (6.115)$$

$$\lambda_i = \frac{\tilde{\lambda}_i}{1-\pi_i} \qquad C_i^2 = \frac{\tilde{C}_i^2 - \pi_i}{1-\pi_i} \tag{6.116}$$

$$\pi_i = \frac{\lambda_{0i}\pi_{0i} + \sum\limits_{j=1}^{M} \lambda_j \hat{p}_{ji} \frac{\pi_{ji}}{(1-\pi_{ji})}}{\lambda_{0i} + \sum\limits_{j=1}^{M} \lambda_j \hat{p}_{ji} \frac{1}{(1-\pi_{ji})}} \tag{6.117}$$

$$\tilde{C}_{dj}^2 = \tilde{\rho}_j^2 \tilde{C}_{Sj}^2 + (1-\tilde{\rho}_j)\tilde{C}_j^2 + (1-\tilde{\rho}_j)\tilde{\rho}_j \qquad C_{dj}^2 = \frac{\tilde{C}_{dj}^2 - \pi_{dj}}{1-\pi_{dj}} \tag{6.118}$$

$$\frac{1}{1+\tilde{C}_i^2} = \frac{\lambda_{0i}}{\lambda_i} \frac{1}{1+\tilde{C}_{0i}^2} + \sum_{j=1}^{M} \frac{\lambda_j \hat{p}_{ji}}{\lambda_i} \frac{1}{1+\tilde{C}_{dji}^2} \tag{6.119}$$

$$\text{with } \hat{p}_{ij} = \begin{cases} p_{ij} \dfrac{1-\pi_{ij}}{1-\pi_{di}} & RS-RD \\ p_{ij} & RS-FD \end{cases} \tag{6.120}$$

Details on the derivations of the above approximating formulae may be found in [KoX89] and [Per94].

The state distribution $p_i(n_j)$ for each queue $i=1,........,M$ is found by solving the GE(λ_i, C_i^2)/GE($\mu_i, \widetilde{C}_{Si}^2$)/$c_i$/$N_i$ queue using the maximum entropy based approximate method of Kouvastos. For a queue of this type, the results are

$$p(n) = p(0)G_n x^{h(n)} y^{f(n)} \qquad n=1,.........N \tag{6.121}$$

Note that unlike the case of closed networks, the queue here is not censored and hence the states may range over the integers 0 to N. For evaluating the state probability using Eq. (6.121), we use the following.

λ = Average arrival rate of the arrival process to the queue

C^2 = SQV of the inter-arrival time of the arrival process to the queue

$$G_n = \prod_{l=K+1}^{m(n)} g(l) \quad J = \max(c,1) \quad h(n) = \max(0, n-J)$$
$$f(n) = \max(0, n-N+1) \quad m(n) = \max(1, \min(c,n)) \tag{6.122}$$

$$g(1) = \begin{cases} \dfrac{\tau c \rho \sigma}{[\sigma(1-\tau)+\tau]} & c > 1 \\ \dfrac{\tau c \rho \sigma}{\tau\rho(1-\sigma)+\sigma} & c = 1 \end{cases} \tag{6.123}$$

$$g(l) = \begin{cases} \dfrac{\tau c \rho + (l-1)\sigma(1-\tau)}{l[\sigma(1-\tau)+\tau]} & l < J \\ \dfrac{\sigma[\tau c \rho + (J-1)\sigma(1-\tau)]}{J[\tau\rho(1-\sigma)+\sigma]} & l = J \end{cases} \qquad l=2,........,J \tag{6.124}$$

with

$$x = \frac{\tau\rho + \sigma(1-\tau)}{\tau\rho(1-\sigma)+\sigma} \quad y = \frac{1}{1-(1-\sigma)x} \quad \sigma = \frac{2}{1+C_S^2} \quad \tau = \frac{2}{1+C^2} \quad \rho = \frac{\lambda}{c\mu}$$

and $p(0)$ may be found by using the normalisation condition that

$$\sum_{n=0}^{N} p(n) = 1 \tag{6.125}$$

The iterations of *Step 3* above are performed until we obtain sufficient convergence in the values of $\widetilde{C}_{di}^2$ and $\widetilde{C}_i^2$.

Note that these equations are very similar to the equations given earlier for the censored queue in the case of closed networks with rejection blocking. The differences between them arise from the fact that the queues are not censored in the case of open networks.

Step 4. All the parameters of interest may now be calculated from the state distribution and flow parameters obtained after the convergence in *Step 3*. These are done in the usual fashion given that the state distributions of the individual queues and of the network are known and the effective flow rate to each queue has also been found.

6.6.3 Transfer Blocking

We have developed [TMB99] a new approximation technique for the analysis of general open networks with multiple servers under transfer blocking mechanism. This approach has been verified to work well in networks with light or medium loading and has been described in this section. In this approach, a hypothetical node is added to handle blocking at the individual queues. The network is decomposed into individual nodes with modified arrival processes. The resulting network is then iteratively solved using flow balance and Maximum Entropy techniques. This approach can handle any general network configuration. Suitable approximations have been included to handle general arrival processes and service times.

It is possible for deadlocks to occur even in an open network of finite capacity queues if the Transfer Blocking mechanism is being used. This is illustrated in the network shown in Figure 6.9. For example, there are several ways in which deadlocks can occur in the network shown in the Figure 6.9. A particularly simple situation is one where Q_1 and Q_3 are filled to their respective capacities and all the servers in Q_1 have blocked jobs waiting to go to Q_3 and vice versa. Such deadlocks can obviously occur in any network of this type, which has feedback, i.e. where a routing loop can be formed. In an actual queueing network, such a deadlock situation cannot be resolved except through external intervention. We assume a similar condition in our analytical approach. Specifically, we assume that any deadlock condition that arises is immediately resolved by moving all the deadlocked jobs simultaneously to their respective destinations.

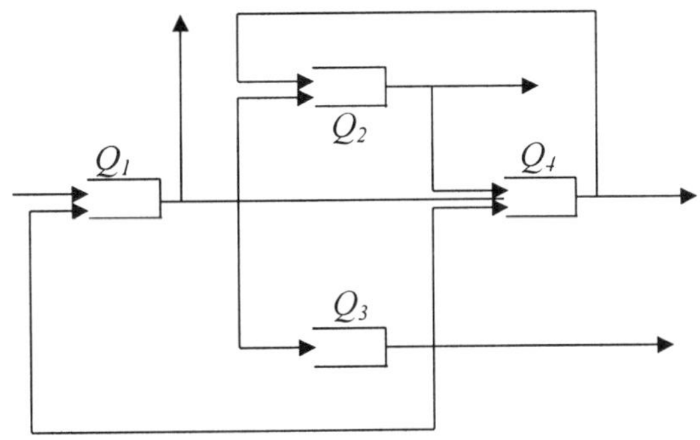

Figure 6.9. Open Queueing Network with Transfer Blocking

For this analysis, the inter-arrival and service time distributions are assumed to be generalised exponential (GE) in nature. A GE distribution is defined through its *mean* and its *squared coefficient of variation* (SQV). A GE distribution has been chosen because of its capability to represent a wide variety of arbitrary distributions, either exactly or approximately. (See Appendix 6.1.) The original network is reconfigured by the addition of a hypothetical node to each incoming stream of all those queues that have finite capacity. This hypothetical node is added to hold the blocked jobs, if any. A blocked job will be routed to this hypothetical node which is modelled as an infinite server queue.

Consider a network of GE/GE/m/N queues which is arbitrarily connected with appropriate routing probabilities which are static in nature. We use Ω to denote this set of queues and assume that the external arrivals (if any) into queue i have a rate λ_{ei} with C_{ei}^2 as the SQV of its inter-arrival times. Let p_{ij} denote the probability that a job finishing service at queue i is routed to queue j. Queue i is assumed to have m_i servers with μ_i as the rate of service and C_{Si}^2 as the SQV of the service time distribution. The buffer capacity of queue i, including the servers, is assumed to be N_i. The steady state probability of the queueing network being in state $\boldsymbol{n}=(n_1,.........n_M)$ is denoted by $\pi(\boldsymbol{n})$ where n_i is the state of queue i. Since the output of a GE/GE/m/N queue may also be approximated by a GE process, we can assume that the product-form solution will approximately hold at equilibrium and therefore, we can write

$$\pi(\boldsymbol{n}) = \prod_{i=1}^{M} \pi_i(n_i) \tag{6.126}$$

where $\pi_i(n_i)$ is the steady-state distribution of queue i which can be found by using the Maximum Entropy based methods of Kouvastos as given in Eq. (6.121) - (6.125).

Since we are assuming finite capacity queues, some jobs will experience blocking where the blocking mechanism being considered here is Transfer Blocking. External arrivals that see a full queue are assumed to be lost, i.e. rejected. Jobs moving from one queue to another, which see a full destination queue, are held at its server in the source queue. This will block the server at the source queue until the destination server has free capacity to accommodate this job. This is because of the nature of the Transfer Blocking mechanism being considered here. Let $PB_j(i)$ be the probability that a job finishing service at queue i and wanting to go to queue j is blocked and let $PBa(j)$ be the probability that job wanting to arrive into queue j is blocked. Let $PBf(i)$ be the probability that a job finishing service at queue i is blocked and let $PBe(i)$ be the probability that an external arrival to queue i is blocked.

We make use of the following three properties of the GE distribution in our analysis. (See Appendix 6.1.) These are -

(i) The departure process from a GE/GE/m/N queue may be approximated by another suitably approximated GE process.
(ii) Bernoulli (random) sampling of these departures, which will correspond to probabilistic (i.e. Markovian) routing will also have a GE distribution
(iii) Combining these sampled processes, which will correspond to the arrival process to a queue in the network, will also have a GE distribution.

However, the process of blocking will further complicate the nature of the arrival and departure processes at the queues. Moreover, the transfer blocking discipline is not work-conserving in nature, and therefore, strictly speaking, the product form distribution cannot be used for the original network. We tackle this problem by making an equivalent work-conserving network by the addition of hypothetical *holding nodes* to the original network. These holding nodes are GE/GE/∞/∞ queues whose GE distribution parameters will correspond to the service times and inter-arrival times corresponding to the blocking delay experienced by a job finishing service and wanting to go to a finite buffer queue. We claim that if we can describe the arrival and service processes to the queues in the equivalent network, then we can solve for the required queueing parameters. For this we use a relaxation method of obtaining a fixed-point solution where the iterations terminate on the convergence of the parameters of interest. We explain this procedure of adding additional holding nodes further using the example network shown in Figure 6.10, where Q_2 and Q_3 are assumed to be ones with finite buffers. This network is expanded to the equivalent work-conserving network by the addition of holding nodes. The resultant expanded network is shown in Figure 6.11. Here we have added holding nodes h_{13}, h_{12}, h_{23} and h_{32} to handle the effects of transfer blocking

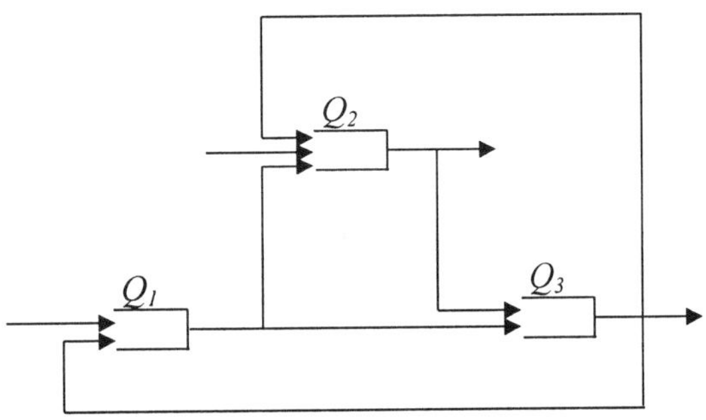

Figure 6.10. Queueing Network with Transfer Blocking (Queues 2 and 3 have finite buffers)

In general, the equivalent work-conserving network is obtained by adding a holding node between every node pair (i,j) in the original network where node j is a queue of finite capacity and $p_{ij}\neq 0$. This holding node would unblock a blocked server by removing the blocked job and holding it in the holding node. The time spent by a blocked job in the holding node should correspond to the time it would be blocked in the original network. Simultaneously, we increase the service time of the other jobs in the queue unblocked in this fashion to reflect the time lost to the jobs behind the blocked job in the actual network. We represent by h_{ij} the holding node added between queues i and j, if any. The set of all holding nodes added in this fashion is denoted by H and the resultant set of all nodes in the equivalent work-conserving network by $H\Omega$. As shown in Figure 6.11 for the original network of Figure 6.10, the holding nodes represent the additional delay caused by the transfer blocking mechanism. We model these holding nodes as GE/GE/∞/∞ queues to introduce delays equivalent to the blocking delays of the blocked jobs.

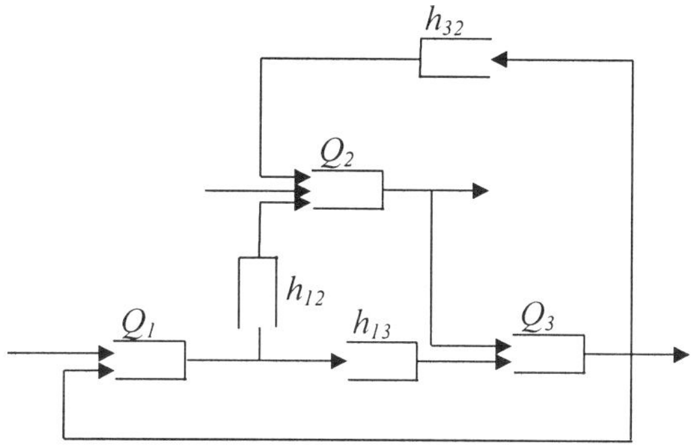

Figure 6.11. Holding Nodes for Jobs along (1,3), (1,2), (2,3) and (3,2)

We then need to obtain the parameters of the arrival and service processes for each of the holding nodes in H. For the holding node h_{ij}, the arrival process is obtained by random (Bernoulli) sampling of the departure process at queue i with probability $PB_j(i)$, the blocking probability for a job leaving queue i for queue j. To obtain the service moments at the holding node h_{ij}, we use the fact that the time for node j to accept the new (previously blocked) job is the minimum of the m_j residual times corresponding to one of the currently served m_j jobs in queue j finishing its service. The moments of the distribution of this time may be approximately calculated from residual time results by using the mean and the SQV of the service time at the blocking node to find the approximate service time distribution. This is fitted as one of exponential, balanced H_2 or E_k kind of distribution depending on whether the SQV is equal to, greater than or less than unity. For this, consider a random variable S which is the minimum of k random variables $s_1,.........,s_k$ whose resultant distribution may then be computed using the distributions of the individual random variables s_i as the ones corresponding to that of the approximated service time distribution. The moments of S may then be computed and may be used to compute the moments of the delay R by using the result [Kle75], [[BrG84] that

$$E\{R^n\} = \frac{E\{S^{n+1}\}}{(n+1)E\{S\}} \tag{6.127}$$

where $E\{R^n\}$ and $E\{S^n\}$ are the respective n^{th} moments of R and S.

Consider once again the blocking scenario of a job getting blocked in attempting to move to queue j from queue i. Specifically consider the network of Figure 6.10 and 6.11 where there are blocked jobs waiting to go to node *3* from both nodes *1* and *2* when the holding nodes h_{13} and h_{23} are both occupied. In this situation, if a job now finishes service at node *3,* then only one of the jobs from either nodes *1* or *2* can move to node *3* and the other job is blocked again. This implies that a "service completion" (corresponding to the destination finishing service to a job) at the holding node might result in one or more rounds of blocking before the job becomes eligible to move to the destination queue *j*. Thus there will be a non-zero blocking probability $PB_j(h_{ij})$ for jobs going from h_{ij} to node *j* and these blocked jobs will return to h_{ij}. This constitutes *immediate feedback* to the node *hij* and needs to be removed following the usual strategy for immediate feedback removal described earlier, for example in Section 6.2. This would further increase the service times at the holding nodes - note that this increase is in effect taking into account the contribution from other nodes feeding into the same blocking node.

Using the above, we can obtain the arrival rates λ_{hij} and the SQV C_{hij}^2 of the inter-arrival times of the holding node h_{ij} as

$$\begin{aligned}\lambda_{hij} &= \lambda_i p_{ij} PB_j(h_{ij}) \\ C_{hij}^2 &= 1 - p_{ij} PB_j(h_{ij}) + p_{ij} PB_j(h_{ij})[\rho_i^2 C_{Si}^2 + (1-\rho_i) C_{Ai}^2 \\ &\qquad + \rho_i (1-\rho_i)]\end{aligned} \tag{6.128}$$

where $\rho_i = \lambda_i / (m_i \mu_i)$.

The mean μ_{hij} and SQV C_{Shij}^2 of the service times of the node h_{ij} also need to be determined. When blocking occurs, all the servers of blocking node will be busy. The distribution of time until any one of the m_i servers finishes service is the residual life of the minimum of m_i GE random variables of rate μ_i and SQV CS_i^2. Let μ_{ri} and C_{Sri}^2 be the mean and SQV of this residual life which can be obtained once again as in Eq. (6.127). The mean and SQV of the service time at h_{ij} may then be approximated by the geometric sum of random variables with mean μ_{ri} and SQV C_{Sri}^2.

$$\mu_{hij} = \mu_{rj}[1 - PB_j(h_{ij})], \quad C_{Shij}^2 = PB_j(h_{ij}) + [1 - PB_j(h_{ij})] C_{Srj}^2 \tag{6.129}$$

We also need to use the following modified routing probabilities.

$$\begin{aligned} p_{ih_{ij}} &= PB_j(i) \\ p_{ij} &= p_{ij} - PB_j(i) \\ p_{h_{ij}k} &= \begin{cases} 1 & k = j \\ 0 & k \neq j, k \in \Omega H \end{cases}\end{aligned} \tag{6.130}$$

The mean number of jobs in the original node (actual network) is the sum of the mean number of jobs in the original node (work-conserving network) and the mean number of jobs in all the holding nodes which emanate from this node. Moreover, the time for which the server will be held in the node in the actual network will be the sum of the times for which it is held at the node in the modified network and the time spent in the holding node.

After obtaining the input queueing parameters for the holding nodes in the modified network, we apply an iterative procedure to solve this network. This is along the lines of a fixed-point solution, which yields approximate steady state solutions for the transfer blocking, open, queueing network being examined.

The iterative solution procedure first obtains the parameters of the arrival process and service times at each node of the expanded network. Each queue is then treated independently using the Maximum Entropy method to find the probabilities $\pi_i(n_i)$, $PB_j(i)$, $PBf(i)$ and $PBa(i)$ as defined earlier. This would require solving a set of coupled non-linear equations. The recursion will finally yield the mean and SQV of the arrival and service processes at the holding nodes. These are then used to modify the statistics of the original network and then the approximate product-form solutions are used to find the overall network parameters

For all the nodes in the expanded network with immediate feedback, we first modify the mean and SQV of their service time in the usual fashion using Eqs. (6.13) and (6.14) as in Section 6.2. The routing probabilities are also adjusted using Eq. (6.15) as given there. The mean and SQV for the arrival processes for nodes $i \in \Omega$ are obtained as follows. The mean arrival rate λ_j into node j is obtained by solving the following set of flow balance equations.

$$\lambda_j[1-PBa(j)] = \lambda_{ej}[1-PBe(j)] + \sum_{i \in \Omega H} \lambda_i[1-PBa(i)]p_{ij}, \; j \in \Omega \quad (6.131)$$

over all the queues in the network. Note that the flow balance equations in Eq. (6.131) take into account both the external flows and their loss due to blocking as well as the internal flows.

The SQV of the inter-arrival times at each node j is obtained as follows. The departures from every node $i \in \Omega H$ are sampled with probability p_{ij} to get the individual streams arriving at node j and the SQV of the combined arrival process at node j is then computed using Eqs. (A6.2) and (A6.3) of Appendix 6.1. For this we need to know the blocking probabilities $PB_j(i)$ from all other nodes i, $PBa(j)$ and $PBe(j)$. These are obtained as follows.

$$PBe(j) = \sum_{n=0}^{m_j-1} \left[\frac{C_{Aji}^2+1}{C_{Aji}^2+C_{Sj}^2}\right]^{m_j-n} \left[\frac{C_{Aji}^2-1}{C_{Aji}^2+1}\right]^{N_j-n} \pi_j(n) + \sum_{n=m_j}^{N_j} \pi_j(n) \left[\frac{C_{Aji}^2-1}{C_{Aji}^2+1}\right]^{N_j-n} \tag{6.132}$$

where $C_{Aji}{}^2$ is the SQV of the inter-arrival times of the jobs from node i to node j obtained using (A6.4) with appropriate substitution.

$$PBa(j) = \frac{\lambda_{ej} PBe(j) + \sum_{i\in\Omega} \lambda_i p_{ij} \dfrac{PB_j(i)}{1-PB_j(i)}}{\lambda_{ej} + \sum_{i\in\Omega} \lambda_i p_{ij} \dfrac{1}{1-PB_j(i)}} \tag{6.133}$$

Note that *PBe(j)* and *PBa(j)* are not equal because arrivals from inside the network into node j are not Poisson in nature.

$$PB_j(i) = \sum_{n=0}^{m_j-1} \left[\frac{C_{Aje}^2+1}{C_{Aje}^2+C_{Sj}^2}\right]^{m_j-n} \left[\frac{C_{Aje}^2-1}{C_{Aje}^2+1}\right]^{N_j-n} \pi_j(n) + \sum_{n=m_j}^{N_j} \pi_j(n) \left[\frac{C_{Aje}^2-1}{C_{Aje}^2+1}\right]^{N_j-n} \tag{6.134}$$

We also need to increase the service time moments of the nodes in Ω to account for the fact that a blocked job is moved to the corresponding holding node. This should have the same effect overall as when the job is actually blocking the server. This will be done as follows

$$\begin{aligned} &\mu_i = [1 - PBf(i)]\mu_i \\ &C_{Si}^2 = PBf(i) + C_{Si}^2[1 - PBf(i)] \qquad i \in \Omega \\ &with\ PBf(i) = \sum_{i\in\Omega H} p_{ij} PB_j(i) \end{aligned} \tag{6.135}$$

The overall algorithm is summarised below.

1. Remove immediate feedback in the network and initialize the following

$PB_j(i)=PBe(j)=PBa(j)=PBf(i)=0 \quad \forall i, j \in \Omega$

2. Expand the network by adding the required holding nodes
3. Obtain the mean inter-arrival times using (6.131) and its SQV as explained therein. Obtain the mean and SQVs of the service times using Eq. (6.135)
4. Using the Maximum Entropy based approach given earlier for open networks, solve for $\pi_i(n_i)$, $i \in \Omega H$
5. Find the blocking probabilities $PB_j(i)$, $PBa(j)$, $PBe(j)$ $\forall i, j \in \Omega H$
6. Do as in Step 3.
7. Repeat from Step 4, if the flow statistics (mean and SQV of the arrival processes at the nodes) have not converged to the desired accuracy.

APPENDIX 6.1: THE GENERALISED EXPONENTIAL DISTRIBUTION

The generalised exponential (GE) distribution is a two-parameter distribution that may be used to approximate a variety of distributions by matching two of their moments. (In contrast, the exponential distribution is a one-moment distribution, which is completely specified by just its mean value.) Its probability density function is given by

$$f_X(x) = \begin{cases} \alpha & x = 0 \\ (1-\alpha)^2 \lambda e^{-\lambda(1-\alpha)x} & x > 0 \end{cases} \tag{A6.1}$$

with $\alpha \in [0,1)$. The mean of this distribution is λ^{-1} and its SQV is $(1+\alpha)/(1-\alpha)$. It may also be viewed as a limiting case of the hyper-geometric distribution where one of the phases is of zero length. For the limiting case of $\alpha=0$, it reduces to the exponential distribution. The GE distribution is useful for modelling processes where the SQV is greater than one. It can also model batch Poisson arrivals with a geometrically distributed batch size. It has also been shown that the GE distribution is a robust two-moment approximation for any service time distribution with SQV greater than one.

Two of the most useful properties of the GE distribution in queueing networks are that it is closed under merging and random sampling. This means that the sum of two or more GE processes (i.e. ones whose inter-arrival times have a GE distribution) is also a GE process. Similarly, random Bernoulli sampling of arrivals from a GE process also leads to a GE process. These properties are similar to the ones that we had seen earlier for Poisson arrival. For pretty much the same reasons, the GE process (like the Poisson process) is easy to handle in the analysis of such networks.

Merging or Superposition of GE Streams

We consider the GE process arising from the merging N GE streams. The ith stream has mean arrival rate λ and has C_i^2 as the SQV of its inter-arrival times. The mean λ_M and SQV C_M^2 of the combined GE process are given by

$$\lambda_M = \sum_{i=1}^{N} \lambda_i \qquad C_M^2 = -1 + \frac{\lambda_M}{\sum_{i=1}^{N} \frac{\lambda_i}{1+C_i^2}} \tag{A6.2}$$

Bernoulli Sampling of GE Streams

If a GE arrival process is sampled with probability p, then the sampled process will also be a GE process. Let the original GE process have mean flow rate λ and SQV (of the inter-arrival times) of C^2. Then the mean and SQV of the sampled process will be given as

$$\lambda_S = p\lambda \qquad C_S^2 = (1-p) + pC^2 \tag{A6.3}$$

Queueing with GE Arrival Streams

Another convenient feature of the GE process in a queueing scenario is that the output from a GE/GE/*m*/*N* queue can also be shown to be approximately GE in nature. Assume that the arrival process to the queue has the mean flow rate λ_{in} with SQV (of the inter-arrival times) of C_{in}^2. Let the mean and SQV of the service times in the queue be μ^{-1} and C_S^2. Defining $\rho=\lambda/(m\mu)$, the mean flow rate of the output process and its SQV will be given by

$$\begin{aligned} \lambda_{out} &= \lambda_{in}(1-P_B) \\ C_{out}^2 &= \rho^2 C_S^2 + (1-\rho)C_{in}^2 + \rho(1-\rho) \end{aligned} \tag{A6.4}$$

with P_B as the probability that a customer arriving to the queue is blocked. This is given by

$$P_B = \sum_{n=0}^{m-1} \pi(n)(1-\tau)^{N-n}\left[\frac{\sigma}{\sigma(1-\tau)+\tau}\right]^{m-n} + \sum_{n=m}^{N} \pi(n)(1-\tau)_{N-n} \tag{A6.5}$$

with $\sigma = \dfrac{2}{1+C_S^2}, \tau = \dfrac{2}{1+C_{in}^2}$ and $\pi(n)$ is the probability that there are n jobs in the queue. Evaluating this probability $\pi(n)$ is a non-trivial task. One approach that may be taken is to use the equivalent state probabilities from an M/M/*m*/*N* for this. The other approach is to use the Maximum Entropy techniques to estimate these probabilities as part of the process of solving the overall queueing network.

Chapter 7

Simulation Techniques for Queues and Queueing Networks

Basic Principles for the Design of Queueing Simulators

In our previous chapters, we have presented various analytical methods for studying the performance of queues and queueing networks. As would be evident from our discussions in those chapters, analytical models of such systems are tractable only if one makes suitable simplifying assumptions. For example, in the study of individual queues, we usually have to resort to such assumptions regarding the nature of the arrival and service processes in order to provide analytical results on the performance of the system. The situation becomes worse for queueing networks where the simple product-form solution holds exactly only under very restrictive assumptions. In some of the methods described for analysing queueing networks, the entire analysis was based on the assumption that even though the product-form expression may not be exactly applicable to the particular network, it still holds as a good approximation. Even with such an assumption (and some even more drastic ones), the analysis of reasonably complicated systems of individual queues or queueing networks becomes rapidly impossible to tackle. In situations where studies of such systems are nevertheless required to be done, there is usually very little choice other than to use simulations and simulation tools to examine the system.

Simulations provide a convenient tool to study complex systems, which cannot be accurately modelled for exact or approximate mathematical analysis. A simulation scenario for a complex system may be set up with as much detail as required - essentially as much detail as it would feasible to handle within the limits of the simulation time that can be spent and the simulation complexity one is prepared to code for. Since analytical modelling usually requires simplifying assumptions of its own, simulations are also useful to provide crosschecks on the results obtained by analysis.

In this chapter, we present the basics of how simulations may be used to study queues and queueing networks. We introduce some of the basic terminology of simulation techniques and review some of the relevant concepts of simulations. We consider in detail how simulators may be constructed to study queues and queueing networks and discuss how one can generate confidence in the results obtained through simulations. The reader can get more details from texts on simulation, such as [BCN96] or [Pay 88]

Given the choice between studying a system through simulations or through analytical modelling, the best idea usually is to use a combination of both the approaches. A system can be simulated to behave very closely to what would actually happen in a real system whereas exact or approximate analysis of the system may only be feasible under very drastic (and sometimes inappropriate) assumptions. However, simulations would generally take a long time to run if one wants to generate results with a high enough degree of confidence. Sensitivity of the various system performance parameters to variations in the values of various input parameters is also easier to study using analytical methods than with simulations.

7.1 Simulation Model of a Real World System

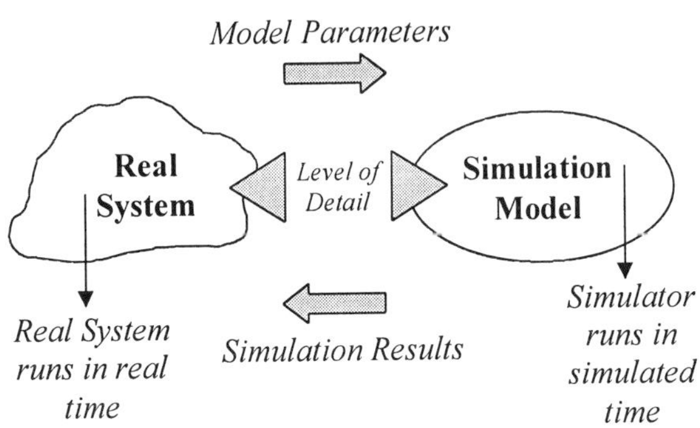

Figure 7.1. Simulation Model of a Real System (Continuous or Discrete States)

The basic ideas behind developing a simulation model for a real system in order to study its behaviour is illustrated through Figure 7.1. This shows the relationship between the real system and its simulation model. A real physical system may have a lot of small details. When one develops a simulation model for it, one tries to make the model as realistic as possible by capturing as many features and details of the real system as feasible. However, it would be extremely fortuitous if we are able to capture all the details of the real life system in our simulator. This may not even be possible in all cases, as the observer may not even know all the fine details of the real system. Even when one has complete knowledge of the real system with all its details, it may not be practical to incorporate all such detail in the simulator. For one, the amount of coding and complexity required to capture all those details in the simulation model may be horrendous. Secondly, a simulator mimicking a real system in such excruciating detail may take an unacceptably long time to run and give results with the desired degree of confidence. One therefore often compromises in building a simulation model of a real system by incorporating only those aspects of the real system, which are deemed to be relevant to the objectives of the study. These *Model Parameters* need to be carefully chosen to satisfy the requirements of the simulation while keeping in mind the complexity of the simulator code and the time required for running it on a computer. During the running of the simulator, one would need to keep track of the values taken on by the parameters that one wants to study through simulations. At the conclusion of the run, the simulator would typically output the moments of the parameters that are being monitored during the simulations. One can also optionally generate a time trace of the some of the selected parameters to see the way these change during the simulation process. Simulations are also sometimes carried out to see if there is any logically inconsistent behaviour in the original system - this, for example, may be an objective if a protocol or algorithm is to be verified through simulations. This is not usually the objective in the simulation of queueing systems. However, simulations of queueing networks may be useful to indicate the different kinds of deadlock conditions that may arise in the real system and suggest possible ways of handling them.

For a given set of values of the model parameters, it is usually not enough to get results just from on single run of the simulator. In order to generate results with some pre-specified degree of confidence, one would usually have to run the simulations several times making sure that the simulation runs are independent and different from each other. One usually modifies the seed in the random number generator, which will be inherent in all simulation code, to ensure this difference. This would ensure that the patterns of events generated by the different simulation runs are sufficiently different even though they are being run from the same set of model parameters.

The real system being simulated will actually run in real time while the simulator will run on a different simulated time, dependent on the simulation code and the system on which the simulation is being run. The way time is handled in a simulator could also be different for different simulators. Since the simulator is being run on a computer, it would necessarily run on a discrete time basis. These discrete time units could be made arbitrarily small if the processing time of the simulator allows this so that it effectively looks like a continuous time simulation. Some real systems, such as discrete-time queues, do have a natural discrete-time operation. In such a system, changes in the system can happen only at discrete time instants and the simulator may effectively use this to simulate events and handle them only at these time instants. A very effective style of simulation only triggers its internal actions when some events affecting the system's state actually occur. This is known as *Discrete Event Simulation* and will be the focus of our attention here.

It is important to emphasise that the simulator is intended to imitate the real system in as much detail as possible or as much detail as is needed to satisfy the objectives of the simulation study. The real system will have its own *entities* with their respective *attributes*, which may interact with each other or have other inter-dependencies as a function of time. We define entities as some object of interest in the real system. For example, in the simulation of a queueing network, the individual queues, the jobs entering/leaving the system and circulating between the queues and the routes followed by them will be some of the important functional entities. The attribute of an entity is some relevant property that we would like to study through simulations. For example, for a particular queue one may like to study attributes such as buffer and server occupancy. For individual jobs, we may be interested in attributes like the time taken to get served in a queue, the time taken to transit a system or whether the job gets lost, say in a finite capacity system. The entities and their attributes would form the *state* of the system as modelled in the simulator. The various *events* that would occur in the simulator (and in the real system) during its operation will affect this state. Selecting the set of events that will be allowed to impact the simulation model and the way they will affect the simulation process will need to be modelled in such a way that they closely imitate what will happen in the real system. As mentioned earlier, the simulation model may be forced to make simplifying assumptions about the real system being studied. However, for the simulation to be meaningful, these must correlate well with the behaviour of the actual system.

The best way to see how a system will perform would undoubtedly be to construct a prototype system and study its behaviour. This is usually not feasible, especially when one wants to study the behaviour of large, complex and costly systems. This is usually the reason why one has to take recourse to either simulations or analysis to obtain results on the behaviour of the system. Given that this is so, whether simulations or analysis is the best strategy is open to debate. Simulators are generally closer to the real system than analytical models as they typically require fewer and less drastic simplifying assumptions than the latter. For example, most analytical results in the area of queueing hinge on the crucial assumptions of Poisson arrivals and possibly exponentially distributed service times. If these assumptions are not acceptable, using other models for the arrival and service processes would be a simple task in a simulator but may be difficult, if not impossible, to handle in the corresponding analytical model. Changing the structure of the model, the rules of the algorithms followed and the values of system variables are sometimes required to see their effects on the system. These changes will be much easier to incorporate in a simulator but will generally be difficult, and may even be impossible to do, in an analytical model. The simulator may also make it easy to study the behaviour of any attribute of any entity in the system by monitoring those during the simulation run. Analytical results for all such variables may not be available which would make it impossible to do a corresponding study using an analytical approach. On the other hand, good simulation models typically take a long time to construct and may also take a long time to run on a computer. This complexity of simulator construction (especially when one wants to construct an efficient simulator using suitable code profiling) and the fact that simulators are generally computationally expensive are the major disadvantages of taking a simulation rather than an analytical approach. Sometimes, the validation of the simulation model and its debugging may also be expensive. Finally, analytical results often allow us to see the relationship between the model variables and the sensitivity of the output parameters to the various inputs. This is not something a simulator can easily do - studying any such change would actually require several simulation runs to see how the system would behave.

Finally, it is perhaps important to point out that building and running a good simulator and getting meaningful results is an art that must be carefully practised. Texts on simulations, including this one, can only point out some of the things that can be done. Much more can usually be achieved by actually building a simulator and carefully examining its operation to see that it is indeed doing things right. Sanity checks using the simulator, where one knows, or can closely guess, how the system should perform in real life are a good way of ensuring this.

Consider a system consisting of a water tank, which is being fed by one or more sources at rates that vary with time. Assume that water also gets taken out of the tank from various outlets at rates that also change with time. We can define the level of water in the tank as the state of the system and can easily construct a simulator to model the real system, i.e. the water tank. For this, we would also need the additional information on how the input flow rates and the output flow rates vary with time. Given these quantities, we can build a simulator quite easily which will tell us what we need to know about the real system, i.e. the way the level of water changes over the day or what would be the level of water at a particular time. A typical output of such a simulator is shown in Figure 7.2.

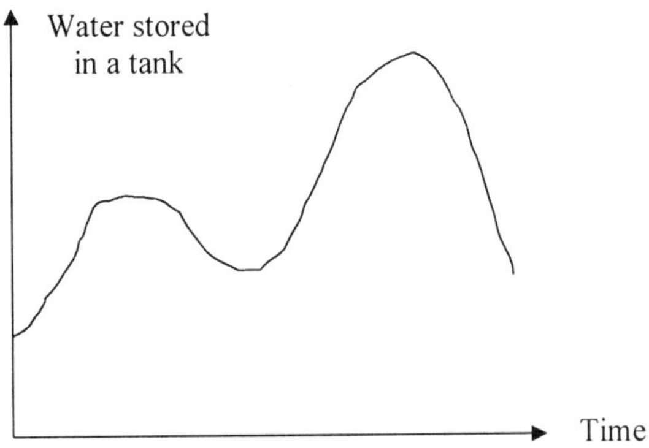

Figure 7.2. Continuous State, Continuous Time Simulations (Level of Water in a Tank)

Note that the system state in this case, i.e. the level of water in the tank, is a continuous variable. Since the state varies continuously with time, it is also a continuous time variable.

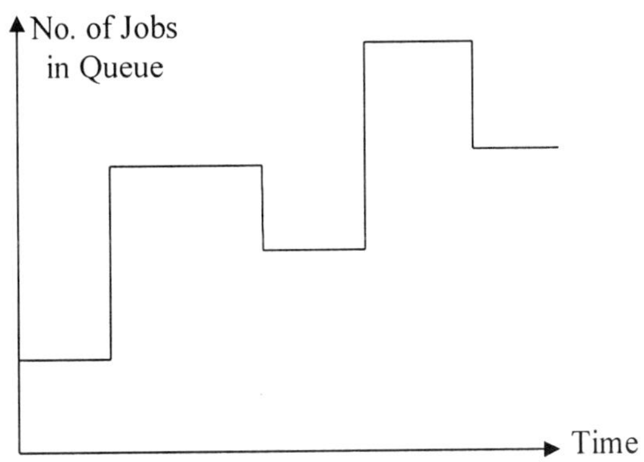

Figure 7.3. Discrete State, Continuous Time System (Number of Jobs in a Queue)

Unlike the example of the water tank shown, a queueing system would have discrete states and would be represented by a discrete variable. An example of this showing the state of a single queue as a function of time is shown in Figure 7.3. Note that for the example shown, the state is a continuous time random variable corresponding to a continuous time queue. If we had instead considered a discrete time queue, then the system state would have been a discrete time variable. In general, whether the system state is continuous or discrete and whether the system is a continuous time or discrete time one would depend on the nature of the system being modelled.

Queueing systems are examples of discrete time systems and we will focus on such systems in our subsequent discussion. In such a system, there will be specific time instants when the model's state will undergo a change. These time instants are sometimes referred to as *simulation times* or *event times*. In between these times, the state of the system does not change. Note that a crude way of simulating a system would be to step it through small intervals of time Δt. We can do this to observe how the system changes from time $t+n\Delta t$ to $t+n\Delta t+\Delta t$ following the rules of how the system state will change over the small time interval Δt. Though this is a possible way to build a simulator, this would not be an efficient way to do so if the system is a discrete state system where the system changes only at specific event times. A *discrete event simulation* approach is a more reasonable way of simulating such a system and will be discussed next

7.2 Discrete Event Simulation

In discrete event simulations, the simulation model only focuses on the event times that may occur as these are the only times when the system is going to change its state. It is not necessary for the simulator to examine (or simulate) the system in between successive event times. This is because the system state will stay unchanged between one event time and the next. This of course implies that the events on the event list should be carefully chosen to represent all the possible events that may affect the state of the system. A continuous simulation of the system in continuous/discrete time is not necessary.

In this case, the simulator always keeps an *event list* (or even multiple event lists which then have to be examined together or are controlled through a *master event list*)) of the events that are currently scheduled to happen and the time instants when they will happen. This is typically organised as a doubly linked list where event $n+1$ keeps one link to its predecessor event n and another to its successor event $n+2$. The events in the list are therefore organised such that event i is listed before event j if event i occurs before event j. At any instant of time during the simulator run, the event which is currently the top most event in the list is the one to be processed next. This event will be processed and the simulator will take all the actions that are required for completing the processing of that event. These actions may change the state of the system and the values and attributes of the various entities affected by this event in the system. The processing of this event may also create new events that have to be scheduled at the appropriate times that they are required to occur. If this happens, then depending on the time of occurrence of these new events, they would be inserted in the appropriate place on the linked list of events. This will be done for all the events generated by the processing of the current event. For example, suppose that the processing of the current event at time t_n creates a new event which is scheduled to occur at time t^*. The simulator will search the event list for the event *i* occurring at event t_i such that t^* falls in the interval (t_i, t_{i+1}). The linked list will then be modified by inserting the new event in between the earlier event *i* and *i+1*. This is illustrated in Figure 7.4 by an example. Note that appropriate special modifications will be needed to handle the cases where the new event is either required to join at the top of the event list (after the event currently being processed) or at the very end of the event list, after the last event that has been currently scheduled.

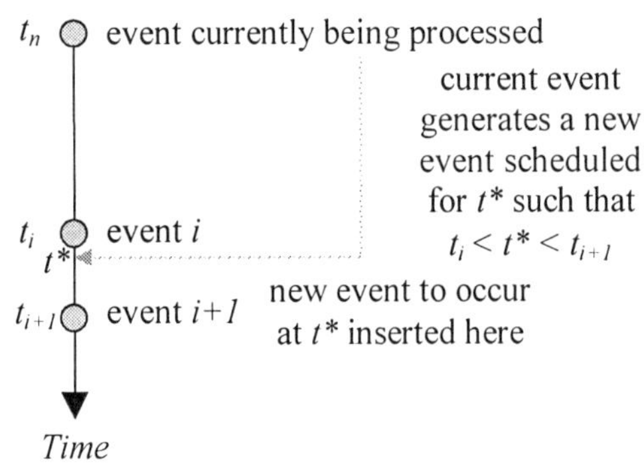

Figure 7.4. Inserting a New Event in the Event List

The simulation process is started with an initialisation procedure. This creates the first few events in the event list and also initialises all the relevant system variables, including those that will be used to record (and later compute) the performance results. Subsequently, the simulation proceeds by processing the event scheduled to occur next which in turn may create additional events that will be placed appropriately in the event list. Once the event at the top of the event list has been fully processed, the simulation time is incremented to the time of occurrence of the next event on the list and this event is then processed. This continues either until the event list becomes empty (no further simulation can be done) or when the simulation termination condition is reached. The simulation termination condition may be specified either as a limit on the simulation time or as a limit on the number of events that will be processed. In the latter case, this may either be a limit on the total number of all events or a limit on the total number of some particular types of events.

The simulation algorithm may be summarised as consisting of the following component procedures. Note that the structure of the simulator is fairly simple consisting of an initialisation procedure, followed by the main simulation loop and the termination procedure. These steps may need to be repeated multiple times with independent runs in order to generate results with the desired degree of confidence.

1. Initialisation Procedure

Setting up the initial event list, initialising the event attributes and variable values internal to the simulator as well as the variables used to monitor the simulation and collect statistics

2. Main Simulation Loop

i. Process current event on top of the event list (i.e. the most recent event - the procedures here will include actions like updating the system state, creating new events and gathering statistics. The actual processing of the event will depend on its type.

ii. Move to the next event on the event list and advance the *simulation clock* to the time of occurrence of this (next) event.

iii. Repeat from (*i*) if the simulation termination conditions stipulated have not been reached

3. Termination Procedure

On termination, process the output parameters to report the required results.

4. Confidence Levels and Confidence Intervals

Repeat 1-3 above with independent simulation runs (i.e. new choice of the seeds for random number generation) until the desired confidence levels for the computed confidence intervals are generated for the simulation results to be considered acceptable.

The procedure for discrete event simulation of a queueing system is best illustrated through an example. For this, we consider the simple queueing network shown in Figure 7.5. This is an open network of two queues with two arrival streams from the outside into the network and probabilistic routing from one queue to another. Various modifications to this simple scenario such as static class-based routing (depending on the external arrival stream) or differences between the priorities of the two streams may also be easily incorporated in this model but have not been considered here.

For the example network of Figure 7.5, the events to be considered in the simulation are as follows -

1. Arrivals from stream 1 (i.e. external arrival stream 1)
2. Arrivals from stream 2 (i.e. external arrival stream 2)
3. Arrivals to queue Q_1 (total arrival stream to Q_1)
4. Arrivals to queue Q_2 (total arrival stream to Q_2)
5. Service completions at (departures from) Q_1
6. Service completions at (departures from) Q_2

These events are inter-related with one event triggering another. We discuss these in detail subsequently.

The initialisation procedure for the event list for this example is very simple. All that we would need to do to start the simulation will be to generate the first arrival from both stream 1 and stream 2 to the network. The subsequent simulations may be designed to follow thereafter as events arising from the processing of these and other subsequent events.

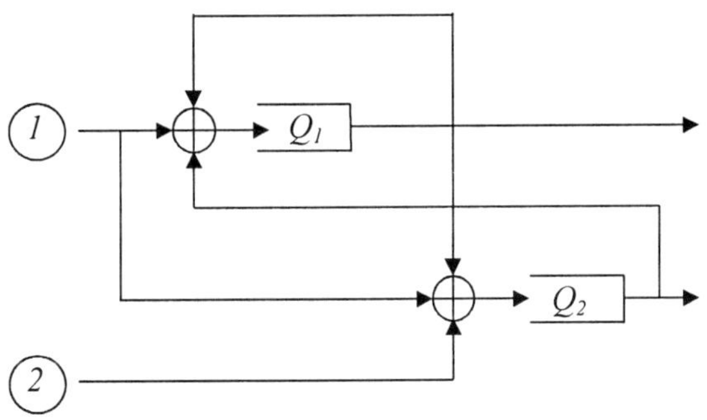

Figure 7.5. An Example of an Open Queueing Network

The following are some of the performance parameters that may be of interest. The ones, which are actually of interest, will have to be measured/recorded during the simulation phase. Once the simulation terminates, average values of these will be computed. Higher moments (e.g. the second moment or the variance) may also be computed. Similar parameters will need to be defined and computed for the actual simulation being done.

1. Number in Q_1 (including/excluding those in service)
2. Number in Q_2 (including/excluding those in service)

3. Waiting time for a job in Q_1 before service or total time spent by a job in Q_1
4. Waiting time for a job in Q_2 before service or total time spent by a job in Q_2
5. Server utilisation at Q_1 or Q_2
6. Sojourn time of a job entering the network from stream *1*
7. Sojourn time of a job entering the network from stream *2*
8. Effective arrival rate entering Q_1
9. Effective arrival rate entering Q_2
10. Departure rate from Q_1 to Q_2 and vice versa
11. Departure rates from Q_1 and Q_2 leaving the network
12. Blocking/loss ratios of the various flows for finite capacity queues

We consider separately the processing required for some of the events in this system, as listed earlier.

External Arrival Event 1

Depending on the nature of the arrival process, generate the next (random) arrival event of this type. This will be scheduled to happen at some time t depending on the nature of the arrival process. Depending on the value of t, this event of next arrival from the arrival stream *1* is placed in the proper position in the current linked list of events. We then generate another random number corresponding to a random choice of whether the arrival is to be routed to Q_1 or Q_2 depending on the routing probabilities given for this. Depending on the value of this random number, the current (external) arrival event is routed to Q_1 or Q_2. This would convert the event to one of an arrival to Q_1 or Q_2 (as the case may be). It is then rescheduled to occur in the same place (i.e. at the same time) as the current event on the event list. Once this is done, move to processing the next event on the event list. (Note that in this case, the next event will occur at the same simulation time instant as the current event but it now becomes an event of another type, i.e. it will be an arrival to either Q_1 or Q_2.

External Arrival Event 2

The processing of this is similar to that of an external arrival event from stream 1, except that the arrival is only routed to Q_2. Therefore, after generating and suitably placing the next event of this type, this event is converted to become an arrival event to Q_2 occurring at the same simulation time instant as before.

Arrival Event to Q_i, $i=1, 2$

If no free server is available, then increment by one the number in the buffer of Q_i. This is done so as to make the arriving job wait in queue for later service. On the other hand, if a server is available then service to job will begin. This involves incrementing by one the number of busy servers at Q_i and then scheduling a departure event corresponding to this job. The departure event is generated at a random time t corresponding to the simulation time when this job will finish service. For this, the random time t is generated as per the service process (or service time distribution) of the server at Q_i. This departure event is placed at its proper position in the event list.

Departure Event from Q_i, $i=1,2$

Depending on the queue from which the departure is taking place and following the routing probabilities for a job leaving that queue, generate a random number to decide which queue the job will be routed to next. If the departure is leaving the network, then we merely need to increment by one the number served by the system. If the departing job is destined for Q_j $j=1, 2$, then convert this event to one of an arrival to Q_j and reschedule it for the same simulation time as the original departure event and place it on the event list.

In addition to the above processing for each of the events (depending on its type), the simulation will also incorporate various counters so that statistical results of the simulation may be computed after termination.

7.3 Collecting and Processing Simulator Outputs for Queues

We had mentioned earlier that we need to monitor and record the values of various variables during the simulation in order to provide the required performance results once the simulation terminates. In this section, we discuss this in more detail in the context of how this will be done in the simulation of single queues or queues in a queueing network

In general, we can divide the variables to be observed for a queue into two classes - *observation-based variables* and *time-weighted variables*. These are discussed separately next.

7.3.1 Observation-Based Variables

Observation based variables are based on individual observations of the queue. Examples of this for a particular queue would be variables like the service time of a particular job or the total time spent by a job in the queue (waiting and in service). For a queueing network, an example of such a variable would be the sojourn time for a job in an open network, i.e. the total time it spends in the network from the time instant when it enters the queue to the time when it finally departs from the system. In a queueing system, variables of this type are usually the ones that we associate with a particular job.

Consider a random variable W of this type. If we knew its actual stochastic properties (i.e. its probability distribution), then we would be able to find its moments of first and higher orders, i.e. typically, its mean $E\{W\}$ and its variance σ_W^2. Consider the situation where K instances of this particular variable are observed as shown in Figure 7.6 where these may be assumed to correspond to the total time spent in the system by the jobs *1, 2, ..., K*.

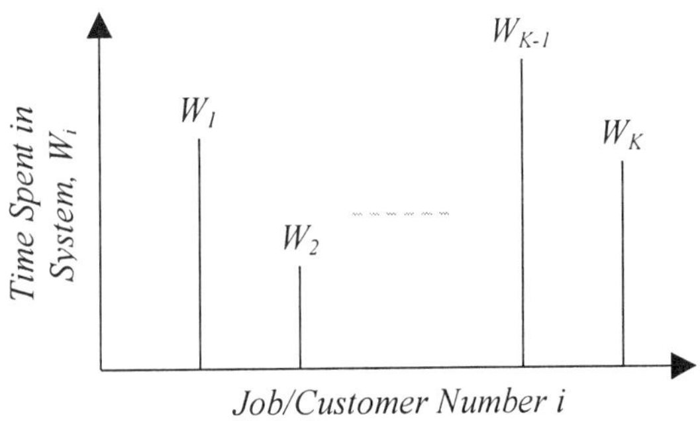

Figure 7.6. Example of an Observation-Based Random Variable (Time Wi Spent in a Queue by the Job i)

If K instances $\{W_1, W_2, \ldots, W_{K-1}, W_K)$ of a particular variable W are actually observed, then its *observed mean* and *observed variance* may be computed as

Observed Mean
$$W_O = \frac{1}{K}\sum_{j=1}^{K} W_j \tag{7.1}$$

Observed Variance
$$\sigma_{WO}^2 = \frac{1}{(K-1)}\sum_{j=1}^{K}(W_j - W_O)^2 \tag{7.2}$$

Other (observed) moments may also be similarly computed. It should be noted that these are merely the observed moments and there is no guarantee that they will indeed be the same or close to the corresponding values of the actual moments that may be obtained had the probability distribution of the random variable W been known. This is where confidence estimation, discussed later, becomes important. The confidence estimation methods will be based on the observed mean and variance of a variable as estimated in a number of independent runs of the simulator.

Finally, it should be pointed out that the observed moments, such as the mean and variance calculated using Eqs. (7.1) and (7.2), are meaningful only when the moments actually exist and can be properly defined. This is possible only when the system is indeed in steady state after all its transients have died out. This has two implications. One is that the computation of these moments is only meaningful in a system that can actually reach steady state. If the queue is overloaded then it is never going to reach steady state and it would not be meaningful to compute the observed moments for such a system. The second implication is that, even for a system which does reach steady state, one must be careful that the observations for W_j $j=1,2......,K$ must be taken only after the initial transient behaviour of the system has died down. This is usually done by prescribing an initial *warm-up time* during which observations are not made. This effectively assumes that the system would reach steady state at the end of the warm-up period. Note that a judicious choice of the warm-up period will have to be made to ensure that this is indeed the case.

7.3.2 Time-Weighted Variables

Time-weighted variables are based on the extent of their actual duration and have to be weighted by the length of this time duration. An example of this would be the number of jobs in the system. This would actually be a function of time $N(t)$, $0 \leq t \leq T$, where T is the total simulation time over which the variable is observed. An example of this is shown in Figure 7.7.

Other examples of such time-weighted variables in the context of queues and queueing networks would be quantities like number of jobs waiting for service in the queue, the number of busy servers in a queue or the total number of jobs in the entire network. During the simulation run, say over the time interval $(0, T)$, one would typically be observing a variable of this type as a function $N(t)$ over the time t. Note that even though $N(t)$ is implied as being a continuous function of time, we really do not need to observe it continuously. It is enough to observe it at the time instants when events actually occur, as we know that the system will not change state in between these time instants. Fortunately, this also agrees very well with the way the simulations are handled in a discrete-event simulator.

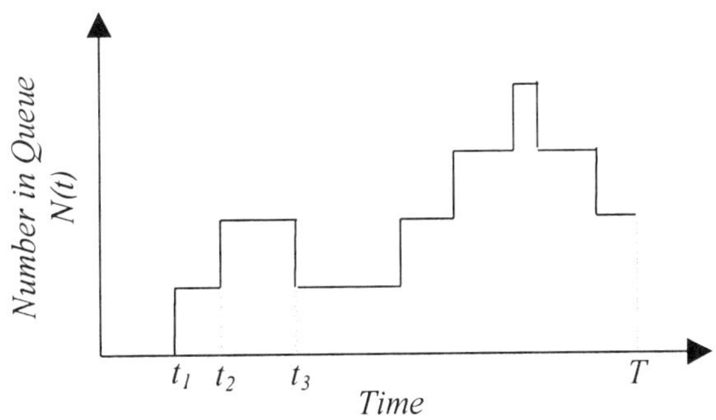

Figure 7.7. Example of an Time-Weighted Random Variable (Number in a Queue as a Function of Time)

The *observed mean* and *observed variance* of the variable *N(t)* may be computed using the following.

Observed Mean $$N_O = \frac{1}{T}\int_{t=0}^{T} N(t)dt \tag{7.3}$$

Observed Variance $$\sigma_{NO}^2 = \frac{1}{T}\left[\int_{t=0}^{T} N^2(t)dt\right] - N_O^2 \tag{7.4}$$

As in the case of observation-based variables, we would need these quantities from independent simulation runs for purposes of confidence estimation. The comments made there about steady state and warm-up intervals are also applicable in this case.

7.4 Estimation of Confidence Intervals and Confidence Levels

It is evident that the point estimation of output variables from the simulator, as described in Section 7.3 from a single simulation run, would be of limited use. If the variance of the variable being observed is large, then the actual "spread" of the random variable would be large. The observed mean of the random variable is basically what one would get from single observation made on it. This can lie anywhere within the range that the random variable can take and there is really no way of saying how close this observed mean would be to the actual mean of the random variable. Ideally, one would like to provide some way of specifying the degree of confidence one can place on such point estimates. This is the reason why estimation of confidence is required for the output of any simulation-based experiment. This is briefly described in this section.

Let W be the random variable being observed where the quantity of interest (from the simulation experiments) is the mean of this random variable. Let $E\{W\}$ be the true mean of the random variable W - this would be what we would be able to calculate if we knew the actual distribution of this random variable. Let W_O be the observed mean of the random variable as obtained by averaging over the observed means obtained from one or more simulation runs. We denote by Δw the *confidence interval* that is estimated for a *confidence level* of α. If confidence estimation is being done, then we would typically be able to make a statement such as "*The true mean $E\{W\}$ of the random variable is expected to lie between $\pm 0.5\Delta w$ of the calculated mean W_O with a probability of α.*" We have illustrated this in Figure 7.8. It should be noted that choosing a range of the type $\pm 0.5\Delta w$ is basically just a convention that we will follow here. Ranges, which are unequal on the two sides of the true mean, may also be considered but are seldom used.

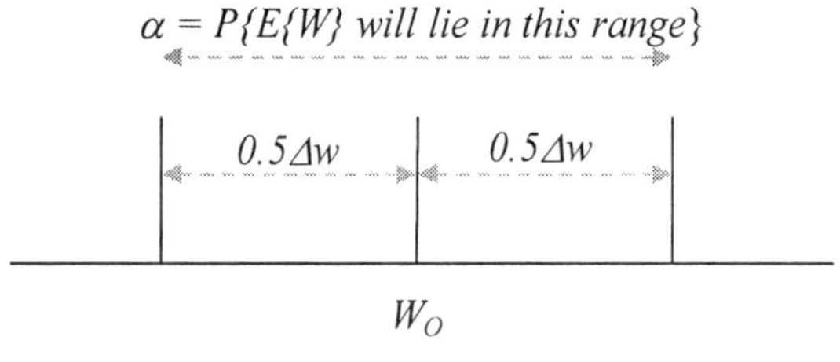

Figure 7.8. Confidence Estimation with Confidence Intervals and Confidence Levels

Note that this is in effect equivalent to saying that, on the average, we would be correct (α x *100*) percent of the time, that $E\{W\}$ will indeed lie in the range ($W_O - 0.5\Delta w$, $W_O + 0.5\Delta w$). The variable α is called the confidence level with Δw the corresponding confidence interval.

Consider a situation where we run our simulation experiment n times to get the sample mean and variance for each run. Then these may be used to get the observed mean and variance as

$$W_O = \frac{1}{n}\sum_{j=1}^{n} W_{Oj} \qquad \sigma_{WO}^2 = \frac{1}{(n-1)}\sum_{j=1}^{n}(W_{Oj} - W_O)^2 \tag{7.5}$$

For a confidence level α, the confidence interval of the observed mean W_O will then be given by

$$\left(W_0 - \left[t_{0.5(1-\alpha),(n-1)} \right] \frac{\sigma_{WO}}{\sqrt{n}}, W_0 + \left[t_{0.5(1-\alpha),(n-1)} \right] \frac{\sigma_{WO}}{\sqrt{n}} \right) \tag{7.6}$$

where $t_{0.5(1-\alpha),(n-1)}$ is the Student's t-parameter for $n-1$ degrees of freedom and σ_{WO} is the standard deviation corresponding to the variance σ_{WO}^2. The value of $t_{0.5(1-\alpha),(n-1)}$ may be found from standard tables. For example, for $\alpha = 0.95$, the values of this parameter for $n=6$ and $n=11$, may be found to be 2.571 and 2.228, respectively.

$$t_{0.025,5} = 2.571 \qquad t_{0.025,10} = 2.228$$

Using this, if for example we do six runs of an experiment, then the actual mean value of the parameter observed would be guaranteed to lie between $(W_O - 1.05\sigma_{WO}, W_O + 1.05\sigma_{WO})$ with 95% probability.

If the confidence level α is given, then confidence estimation would estimate the range of the confidence interval $\pm 0.5\Delta w$ around the observed mean within which we would expect the actual mean value to lie. We can summarise this as follows.

i. Observe the required variable W for n runs

ii. From these n observations, compute the observed mean W_O and the observed variance σ_{WO}^2

iii. Obtain the required Student's t-parameter $t_{0.5(1-\alpha),(n-1)}$ for a confidence level α

iv Use Eq. (7.6) to compute the range of the confidence interval within which we would expect the actual mean $E\{W\}$ to lie with (α x 100) percent confidence.

Note that if the confidence interval obtained in this fashion is not tight enough to suit the experimental requirements then it can be made tighter by doing more experimental runs. This will change the value of n in the steps given above and will lead to tighter confidence intervals. The usual procedure followed is to keep doing simulation runs of the experiment until the confidence interval for each of the required result variables become small enough to be acceptable for the specified level of confidence. Note that, given the confidence level, it is not possible to specify in advance the number of simulation runs that will be required to get the required tightness of the confidence interval for the observed variable. To complicate matters further, if a number of variables are being observed, then the confidence intervals obtained may be different for each of them. The simulation runs will have to be continued until the required confidence intervals of all the variables are individually satisfied, i.e. the confidence interval of each of the variables are equal or smaller than the corresponding required level.

While reporting the results of a simulation experiment, it would be desirable to mention the confidence estimate (confidence level and confidence interval) associated with each of the observed variables. Typically, reporting results in this fashion would be more meaningful than giving just the mean results.

7.5 Transient Behaviour and the Warm-up Interval

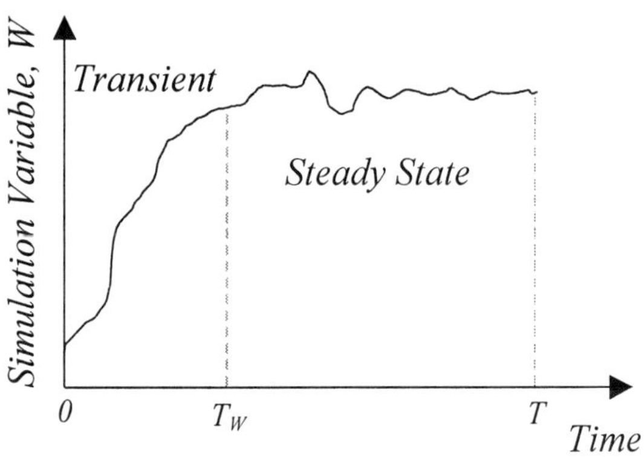

Figure 7.9. Transient Behaviour in a Simulation Run

In Figure 7.9, we have illustrated the typical behaviour of a simulation parameter W during a simulation run carried out over a (simulation) time interval $(0, T)$. Note that the parameter W being observed goes through a transient stage when the simulations are first started but eventually reaches steady state values. (We obviously assume that the simulations are being done with input parameters such that the system will be able to reach steady state.) Though W has been shown to be a continuous random variable, such as the time spent in system, discrete variables like the system state will also show analogous behaviour - except that they will take discrete values. Since the simulation output is typically required under equilibrium conditions, we really should monitor and record the values of W (for calculating the observed mean and variance for the simulation run) only during the time when the system has reached equilibrium. This is done by specifying an initial *warm-up interval*, which should be at least as large as the transient duration, during which system statistics are not recorded.

In Figure 7.9, looking at the time trace of $W(t)$, we have decided to label the interval $0 \leq t < T_W$, as the transient interval and the interval $T_W \leq t < T$ as the one during which the simulation has reached equilibrium conditions. The warm-up interval, mentioned earlier, should be at least as large as $(0, T_W)$, if not larger. Deciding a reasonable value for T_W seems obvious from the plot for $W(t)$ shown in Figure 7.9 but in practice would be hard to do automatically inside a computer program. The reason for this lies in the fact that the effect of transients and the transient interval are both going to be highly dependent on the actual simulation model and even possibly on the particular variable being observed.

We cannot really give any general guidelines on how the warm-up interval, covering the effects of the transients, should be chosen. One possibility is that we choose the overall simulation interval $(0, T)$ to be large enough so that the effects of the transients will be negligible and may be ignored. The other option is to actually observe the simulation experiment for a few independent runs and base a suitable choice of the warm-up period based on the observations made for these runs. Subsequent runs may be done using this value of the warm-up period (during which data is not collected) for calculating the observed means and variances.

7.6 Data Collection in Steady State Conditions

We briefly discuss in this section some of the common methods used to gather steady state observation data during a simulation run. The problem specifically considered is the way one can get n independent simulation runs over which the output parameters may be observed and estimated so that the results from the runs may be used to get a suitable confidence estimate. The three techniques commonly proposed for this are *(i) The Subinterval Method*, *(ii) The Regenerative Method* and *(iii) The Replication Method*. In our earlier discussions, we had in a way implied the use of the replication method as that is indeed the one that is most commonly used.

7.6.1 The Subinterval Method or the Method of Batch Means

This method really involves only one very long simulation run which is suitably subdivided into an initial transient period and n batches. Each of these batches is then treated as an independent run of the simulation experiment while no observations are made during the transient period which is treated as the warm-up interval. An example of this has been shown in Figure 7.10. Note that one would normally choose batch intervals of equal size but this is not mandatory. One can choose batch intervals of different lengths if there is a reasonable logic on which this may be based.

Choosing a large batch interval size would effectively lead to independent batches and hence, independent runs of the simulation as required for confidence estimation purposes. One advantage of this method is that only one transient (warm-up) interval needs to be accounted for and discounted and removed during the process of recording observations. Since the simulation would tend to run for a long simulation time in this approach, this would also tend to reduce the ill effects, if any, of not properly removing the transient period during the warm-up interval.

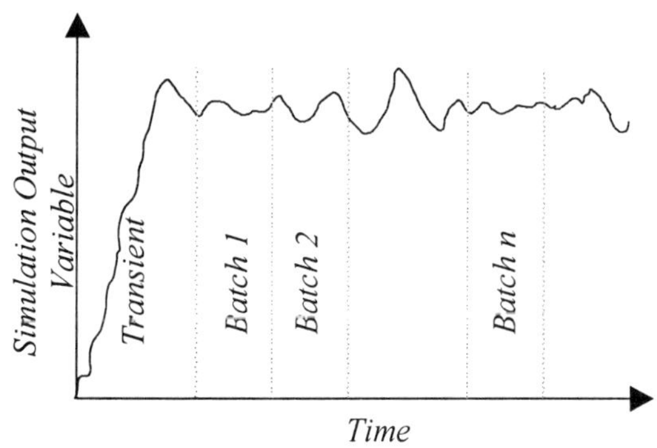

Figure 7.10. The Subinterval Method or the Method of Batch Means

One disadvantage of this approach is that the batches, as illustrated in Figure 7.10, may not really be independent. Statistically, one can make the observation that high values of an observed parameter will tend to follow high values and low values will tend to follow low values. This would be especially true if the batch sizes are not large enough and will lead to significant correlation between successive batches. Since the confidence estimation measures are based on the independence of the individual simulation runs, this lack of independence between successive batches, if sufficiently serious, may have a serious impact on the accuracy of the confidence estimation procedures.

7.6.2 The Regenerative Method

This method is basically intended to reduce the problem of correlation between the batches that one may encounter in the Subinterval Method of Section 7.6.1. We still use one long run as before but select an appropriately identified state of the system as the *regenerative state* and the time instants when this occurs as the *regenerative points*. The batches start and end at these regenerative points once steady state has been reached. This method is illustrated in Figure 7.11.

Choosing the regenerative state appropriately may not be as difficult as it may first appear. Consider the variable N representing the number in a queue, which is being simulated. This will typically build up from a *zero* value but, if the system is stable, then it will also eventually return to a *zero* value. This will happen repeatedly with the system state fluctuating between positive, non-zero integer values in between the zero-value points. The zero-value points are interesting because once the system reaches this state, its future evolution does not depend on how it actually reached this state. This behaviour of the system implies that the way the system evolves in one *idle-to-idle* cycle of this type will not depend on previous cycles of this type and will not affect future cycles of this type either. Since this is the case, we can indeed choose these *idle-to-idle cycles* as the independent batches suggested in the earlier description and illustrated in Figure 7.11.

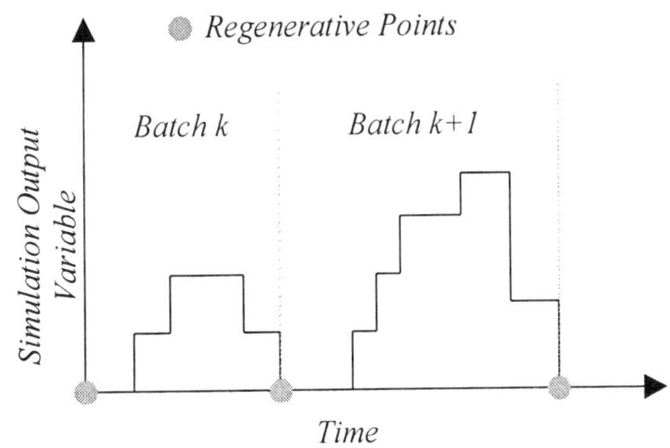

Figure 7.11. The Regenerative Method of Choosing Independent Batches

This method is qualitatively better at ensuring independence between successive batches than the earlier *Subinterval Method*. Apart from the fact that one must correctly identify the regenerative states, this method does have a few problems that should be considered. Firstly, the time interval between regenerative points may be long - if this happens then one will need to run the simulation for a long time (in terms of the simulation time) in order to get a sufficient number of independent batches. The adverse effect of this, if it happens, will also be that only a few batches may be obtainable even in a long simulation run. If this is the case then it will adversely affect the computation of the confidence intervals for confidence estimation purposes. One other problem is that the batch sizes and their duration will themselves be random in nature. It may be necessary to account for this in our computations.

7.6.3 The Replication Method

This method is the one most popularly used and had been implied in a way in our earlier descriptions. This was basically the suggested technique for getting n independent runs of the simulation experiment by running the simulator n times with different initial random seeds for the simulator's random number generator. We have illustrated an example of this in Figure 7.12 where the independent runs for the two batches, batch k and batch $k+1$ are shown with their own transient (warm up) intervals and the time intervals during which the system reaches steady state in these two runs.

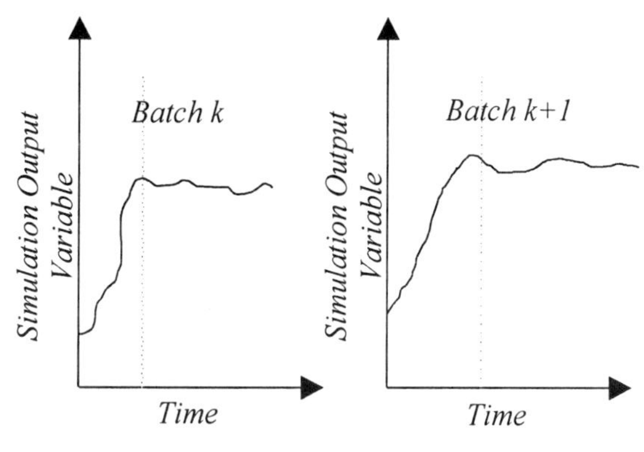

Figure 7.12. The Replication Method

In this case, n independent replications of the simulation run are done. In each case, an appropriately selected transient (warm up) period is removed from each run before the data collection is started. For the observed intervals after the warm up period (i.e. the intervals during which the system reaches steady state), data is collected and processed for the point estimates of the variables being observed and for their subsequent confidence estimation.

The n simulation runs must be truly independent. In practice, this may generally be satisfactorily achieved by selecting different seeds for starting the random number generator used by the simulator. In order to achieve independent replications in this fashion, one should ensure the use of a good random number generator and sufficiently statistically different starting seeds for each run. This method does require separate warm up periods to be estimated and removed from each run to eliminate the effects of transients. Apart from this, the replication method is simple and easy to use and is the method typically used to generate multiple simulation runs of the simulation experiment for confidence estimation.

APPENDIX 7.1 GENERATING RANDOM NUMBERS

In most computer systems, one can find a random number generator (usually a function called *rand*) which may be called to generate uniformly distributed real valued random numbers between *0* and *1*. Typical usage of this generator is to specify a seed first before the random number generator is used. The seed is usually any randomly chosen large integer with the number of digits corresponding to the largest size integer that may be represented on that computer. Once the seed has been specified, calling the random number generator function repeatedly will generate a sequence of uniformly distributed random numbers. Note that the same sequence of random numbers is always generated if the seed value is the same. However, specifying a different seed will generate a different sequence of uniformly distributed random numbers.

The random number generator that typically comes with the system is called *ran(iseed)* or *rand(iseed)*. The usual method of calling this function would be to say

$$x = rand(iseed)$$

which would give x as the next random number generated and will also automatically update *iseed* for the next time the function is called. The random number generators supplied with most computer systems are usually quite poor in guaranteeing true randomness of the sequence of random numbers they generate. It is usually a much better idea to write ones own or to use a random number generating routine from a good reference source such as *Numerical Recipes* [PTV92].

Since most system supplied or referenced random number generators only generate a sequence of random numbers uniformly distributed between *0* and *1* (i.e. random number X, $0<X<1$), we still have the problem of using this random number generator to generate random numbers with other distributions. (Some systems may provide means of generating Gaussian random variables or exponentially distributed random variables.)

The easiest way to do this is by using the so-called *Transformation Method*. Suppose we have a random number generator generating uniformly distributed random numbers X between *0* and *1*. If we are actually required to generate a random variable Y with *cumulative distribution function F(y)*, then we can do that by using the following transformation to generate the required Y from X.

$$Y = F^{-1}(X) \tag{A7.1}$$

This would mean that we first get X and then apply the transformation of Eq. (A7.1) to get Y. For example, if we want to generate the random variable Y to have an exponential distribution (cumulative distribution function of $e^{-\lambda Y}$) i.e. for Poisson arrival processes or exponentially distributed service times) of mean λ^{-1} then the required transformation needed is

$$Y = -\frac{\ln(X)}{\lambda} \tag{A7.2}$$

The transformation method is simple but suffers from the inherent disadvantage that one cannot really use it unless the function F representing the cumulative distribution function of the target random variable can be inverted, i.e. its inverse function $F^{-1}(X)$ may be evaluated either analytically or numerically.

The *Rejection Method* offers a more general approach to generate random variables with a required distribution. The only requirement is that the target random variable must have a *probability density function* $f_Y(y)$, which is known and computable. (Note that $f_Y(y)dy$ actually gives the probability that the random variable Y falls between y and $y+dy$.). It does not require that the cumulative distribution function be readily computable or that it be in a form where its inverse can be analytically or numerically computed, as required in the transformation method. Discussion of this method and of computer programs implementing it for various target distributions may be found in the literature, e.g. in [PTV92].

References

[AkK93] Akimaru, H. and Kawashima, K. 1993. *Teletraffic Theory and Applications*, Springer-Verlag.

[Aky89] Akyildiz, I.F. 1989. Product-Form Approximations for Queueing Networks with Multiple Servers and Blocking, *IEEE Trans. on Computers*, vol. 38, no. 1, pp 99-113, January 1989.

[BCM75] Baskett, F., Chandy, K. M., Muntz, R. and Palacios, F.G. 1975. Open, Closed and Mixed Networks of Queues with Different Classes of Customers, *J. of the ACM*, vol. 22, no. 2, pp 248-260, April 1975.

[BCN96] Banks, J., Carson, J.S., Nelson, B.L. 1996. *Discrete-Event System Simulation*, Prentice-Hall Inc.

[BeG92] Berstsekas, D., and Gallager, G. 1992. *Data Networks*, Prentice-Hall.

[BrG84] Bronshtein, O. and Gertsbakh, I.B. 1984, An Open Exponential Queueing Network with Limited Waiting Spaces and Losses: A Method of Approximate Analysis, *Perfromance Evaluation*, vol. 4, pp 31-43, 1984.

[BrK93] Bruneel, H., and Kim, Byung G. 1993. *Discrete-Time Models for Communication Systems Including ATM*, Kluwer Academic Publishers.

[CHW75] Chandy, K.M., Herzog, U. and Woo, L. 1975. Parametric Analysis of Queueing Networks, *IBM Journal of Research and Development*, Jan. 1975, pp 19-36.

[Fel65] Feller, W., 1965. *An Introduction to Probability Theory and its Applications, Vol. II*, New York: John Wiley & Sons.

[Hay84] Hayes, J. 1984. *Modeling and Analysis of Computer Communication Networks*, New York: Plenum Publishing Corporation.

[KeS87] Kerbache, L. and Smith J.M. 1987. The Generalized Expansion Method for Open Finite Queueing Networks, *European Journal of Operational Research*, 32 (1987), pp 448-461.

[KiA89] Kim, C. and Agarwala, A.K. 1989. Analysis of the Fork-Join Queue, *IEEE Trans. on Computers*, vol. 38, no. 2, pp 83-121, February 1989.

[Kle75] Kleinrock, L. 1975. *Queueing Systems, Vol. 1*, John Wiley & Sons.

[KoX89] Kouvastos, D.D. and Xenios, N.P. 1989, MEM for Arbitrary Queueing Networks with Multiple General Servers and Repetitive Service Blocking, *Performance Evaluation*, vol. 10, pp 169-195, 1989.

[LiP91] Liu, Y.C. and Perros, H.G., 1991. A Decomposition Procedure for the Analysis of a Closed Fork/Join Queueing System, *IEEE Trans. on Computers*, vol. 40, no. 3, pp 365-371, March 1991.

[LZS84] Lazowska, E.D., Zahorjan, J., Scott Graham G., and Sevcik, K.C. 1984. *Quantitative System Performance*, Prentice-Hall Inc.

[Mol89] Molle, M.K. 1989. *Fundamentals of Performance Modelling*, Macmillan.

[NeT88] Nelson, R. and Tantawi, A.N. 1988. Approximation Analysis of Fork/Join Synchronization in Parallel Queues, *IEEE Trans. on Computers*, vol. 37, no. 6, pp 739-743, June 1988

[Onu90] Onuvral, R.O. 1990. Survey of Closed Queueing Networks with Blocking, *ACM Computing Surveys*, vol. 22, no. 2, pp 83-121, June 1990.

[Pap65] Papoulis, A. 1965. *Probability, Random Variables and Stochastic Processes*, McGraw-Hill.

[Pay88] Payne, James A. 1988. *Introduction to Simulation*, McGraw-Hill Inc.

[Per94] Perros, H. G. 1994. *Queueing Networks with Blocking*, Oxford University Press.

[Per89] Perros, H.G. 1989. A Bibliography of Papers on Queueing Networks with Finite Capacity, *Performance Evaluation*, vol. 10, pp 255-260, 1989.

[PTV92] Press, W.H., Teukolsky, S.A., Vellerling, W.T. and Flannery, B.P. 1992. *Numerical Recipes: The Art of Scientific Computing*, Cambridge University Press.

[ReL80] Reiser, M. and Lavenberg , S.S. 1980. Mean Value Analysis of Closed Multichain Queueing Networks, *J. of the ACM*, vol. 27, no. 2, pp 313-322, April 1980.

[Ros95] Ross, K. W. 1995. *Multiservice Loss Models for Broadband Telecommunication Networks*, Springer-Verlag.

[Tak94] Takagi, H. 1994. *Queueing Analysis, Vol. 2: Finite Systems*, North-Holland.

[TMB99] Tahilramani, H. Manjunath, D. and Bose, S.K. 1999, Approximate Analysis of Open Network of GE/GE/m/N Queues with Transfer Blocking, *MASCOTS'99*, University of Maryland, October 24-28, 1999, pp 164-172

[Wal88] Walrand, J. 1988. *An Introduction to Queueing Networks*, Prentice-Hall, Inc..

[Whi83] Whitt, W. 1983. The Queueing Network Analyzer, *Bell System Technical Journal*, vol. 62, no. 9, pp 2779-2815, Nov. 1983.

[Wol89] Wolff, R.W. 1989. *Stochastic Modeling and the Theory of Queues*, Prentice-Hall Inc..

[Woo94] Woodward, M. E. 1994. *Communication and Computer Networks*, IEEE Computer Society Press.

Index